Destructive Production, Agroecology and Schools of Agroecology in Brazil

Studies in Critical Social Sciences Book Series

Haymarket Books is proud to be working with Brill Academic Publishers (www.brill.nl) to republish the *Studies in Critical Social Sciences* book series in paperback editions. This peer-reviewed book series offers insights into our current reality by exploring the content and consequences of power relationships under capitalism, and by considering the spaces of opposition and resistance to these changes that have been defining our new age. Our full catalog of *SCSS* volumes can be viewed at https://www.haymarketbooks.org/series_collections/4-studies-in-critical-social-sciences.

Series Editor
David Fasenfest (York University, Canada)

Editorial Board
Eduardo Bonilla-Silva (Duke University)
Chris Chase-Dunn (University of California–Riverside)
William Carroll (University of Victoria)
Raewyn Connell (University of Sydney)
Kimberlé W. Crenshaw (University of California–LA and Columbia University)
Heidi Gottfried (Wayne State University)
Alfredo Saad-Filho (Queen's University, Belfast)
Chizuko Ueno (University of Tokyo)
Sylvia Walby (Royal Holloway, University of London)
Raju Das (York University)

Destructive Production, Agroecology and Schools of Agroecology in Brazil

Henrique Tahan Novaes and João Henrique Souza Pires

Translated by
Diogo de Lima Saraiva and Semaj Moore

Haymarket Books
Chicago, IL

First published in 2024 by Brill Academic Publishers, The Netherlands
© 2024 Koninklijke Brill NV, Leiden, The Netherlands

Published in paperback in 2025 by
Haymarket Books
P.O. Box 180165
Chicago, IL 60618
773-583-7884
www.haymarketbooks.org

ISBN: 979-8-88890-551-7

Distributed to the trade in the US through Consortium Book Sales and Distribution (www.cbsd.com) and internationally through Ingram Publisher Services International (www.ingramcontent.com).

This book was published with the generous support of Lannan Foundation, Wallace Action Fund, and the Marguerite Casey Foundation.

Special discounts are available for bulk purchases by organizations and institutions. Please call 773-583-7884 or email info@haymarketbooks.org for more information.

Cover design by Jamie Kerry and Ragina Johnson.

Printed in the United States.

Library of Congress Cataloging-in-Publication data is available.

Contents

List of Figures and Tables IX
Acronyms and Abbreviations X

Introduction 1

1 The Current State of "Primitive Accumulation": Land Theft and Enclosures in the Twentieth and Twenty-First Centuries 11
   1 Introduction 11
   2 So-called Primitive Accumulation 13
   3 The Enclosure and Theft of Land in Brazil in the 20th and 21st Centuries 21
   4 The Commodification of All Spheres of Life and Struggles of Resistance against "Primitive Accumulation" 25

2 Destructive Production and Agroecology 31
   1 Introduction 31
   2 Green Revolution or Green Con? The Advancement of Destructive Forces in the Countryside 33
   3 Patents as a New Form of Colonialism 37
   4 Rescue of Historical Experiences of Alternative Agriculture: Clues to the Understanding of Agroecology 40
   5 The Concept of Agroecology and the Need for an Agroecological Transition 43
   6 The Technical Assistance Required for Agroecology 49
   7 The Heterogeneity of Agroecology: From Market Niches to Systemic Rupture 51

3 "Sustainable Development", Agroecology and Ecosocialism 56
   1 Introduction 56
   2 "Sustainable Development" and Its Limits 58
   3 Technological Dependence and Neo-colonial Reversal: Effects on Commodity Exports and Brazil's Role in the International Division of Labour 64
   4 The Contributions of Michael Lowy and István Mészáros to the Ecosocialist Transition 68

4 "Green Revolution" in Brazil, Rural Extension and the Fight to Establish MST 77
   1 Introduction 77
   2 The "Green Revolution" in Brazil 78
   3 Importing the Model of Rural Extension 82
   4 The Fight to Establish MST 92

5 Perspectives and Dimensions of Agroecology 101
   1 Introduction 101
   2 The Perspective of the North-American Thought: For Sustainable Processes in Agriculture 105
   3 Resistance and Existence: Agroecology in Spanish Thought 108
   4 Agroecology in the Brazilian Scenario 113
   5 Agroecology from the Perspective of MST and Social Struggles 117
   6 The Dimensions of Agroecology 127
   7 The Holistic Approach 128
   8 The Participative Approach 134
   9 Educational Dimension 137

6 Transnational Corporations, the MST's Agroecological Agenda and Agroecology Schools 140
   1 Introduction 140
   2 Monster Corporations and the Fetishism of the "Green Revolution" 141
   3 Fights for Agroecology and the MST's Agroecological Agenda 145
   4 Class and Gender Issues in Agroecological Struggles 146
   5 The People's Agrarian Reform and the Construction of the Revolution in Latin America 148
   6 Educational Resistance: The Experiences of the MST Agroecology Centres 152

7 The Political Economy of the "Green Revolution", Agroecology and the MST Agroecology Schools 164
   1 Introduction 164
   2 Capital's Agriculture Campaign and Destructive Production: The Political Economy of the "Green Revolution" 164
   3 Agroecology in the MST: Beyond the Green Agenda 169
   4 The Creation of the Agroecology Schools 173
   5 Agroecology in the Curriculum of the MST Vocational Schools 175
   6 Final Thoughts 176

8 Cooperation and Workers' Cooperatives of MST   178
   1 Introduction   178
   2 From the Coffee Complex to the Expansion of Agribusiness in the State of São Paulo   178
   3 Conceiving Cooperation in the MST   183
   4 Cooperation of the São Paulo MST in Face of the Capitalist State   188
   5 Cooperation and Cooperativism in the MST of São Paulo   190
   6 Final Considerations: Islands/Settlements Surrounded by a Green Sea of Sugarcane and Eucalyptus   199

9 The Rescue of Labour School Principles by the MST and Their Influence on Agroecology Schools   200
   1 Introduction   200
   2 The Re-release of Books from the First Phase of Soviet Pedagogy by the MST   201
   3 The River that Divides the Pedagogies of Capital and the Pedagogies of Labour   202
   4 Fundamentals of the Labour School   203
      4.1 *The Single School of Labour*   204
      4.2 *Polytechnic Schools*   206
      4.3 *Self-direction*   207
      4.4 *Thematic Complexes*   207
   5 The Bureaucratization of the Russian Revolution and Its Educational Contingencies   209
   6 Experimenting with the Principles of the Labour School in the MST's Agroecology Schools   210
   7 The Urgency of an Education beyond Capital   211

Conclusions   215

Bibliography   219
Index   250

Figures and Tables

Figures

5.1 Approaches and dimensions of agroecology   127

Tables

6.1 Pedagogical principles of the José Gomes da Silva school   155
6.2 Role of work teams in José Gomes da Silva School   156
6.3 Description of the educational times of the "earth revolutionaries" class   158

Acronyms and Abbreviations

AACT	Antônio Companheiro Tavares Settlement
ABA	Brazilian Association of Agroecology
ABAG	Brazilian Association for Agribusiness
ABCAR	Brazilian Association for Credit and Rural Assistance
ABRA	Brazilian Association for Agrarian Reform
ACAR	Credit Assistance and Rural Assistance
ANA	National Liason for Agroecology
ATEMIS	Association of Workers in Education and Production in Agroecology Milton Santos
ATER	Technical Assistance and Rural Extension
BACEN	Central Bank of Brazil
BB	Bank of Brazil
BM	Mundial Bank
BNCC	Nation Corporate Credit Bank
CEDI	Ecumenical Center for Documentation and Information
CEPAL	Economic Commission for Latin America
CIMA	Center for Irradiation and Management of Agrobiodiversity
CLOC	Latin American Coordinator for Rural Organizations
CNA	National Council of Agriculture
CNBB	National Conference of Bishops of Brazil
COMPATER	National Commission of Agricultural Research and Technical Assistance and Rural Extension
CONCRAB	Agrarian Reform Cooperative of Brazil
CONTAG	Brazilian Confederation of Workers in Agriculture
CPA	Agricultural Production Cooperative
CPP	Political Pedagogical Coordination
CPT	Pastoral Commission of the Land
CRB	Brazilian Rural Confederation
CTNBIO	National Technical Commission for Biosecurity
CUT	The Only Workers Center
EJA	Education for Youth and Adults
EJGS	José Gomes da Silva School
ELAA	Latin American School of Agroecology
EMATER	Technical Assistance and Rural Extension Company
EMBRAPA	Brazilian Agricultural Research Agency
EMBRATER	Brazilian Entity for Technical Assistance and Rural Extension
EMS	Milton Santos School

ENERA	National Meeting of Educators of Agrarian Reform
ETA	Brazil Technical Bureau – United States
FAO	Food and Agriculture Organization of the United Nations
FASE	Federation of Organizations for Social and Educational Assistance
FHC	Fernando Henrique Cardoso
FMI	International Monetary Fund
FNDEP	National Forum in Defense of Public Schools
FNMA	National Fund for the Environment
GT-RA	Working Group in Support of Agrarian Reform
IALA	Latin American Institutes of Agroecology
IBASE	Brazilian Institute for Socioeconomic Analysis
IBC	Brazilian Coffee Institute
IBGE	Brazilian Institute of Geography and Statistics
IFPR	Federal Institute of Paraná
INCRA	National Institute of Colonization and Agrarian Reform
ISEC	Institute of Sociology and Peasant Studies
ITEPA	Institute of Technical Education and Research for Agrarian Reform
ITERRA	Institute for the Training and Research on Agrarian Reform
MAB	Brazilian Movement of Dam-Affected People
MAPA	Ministry of Agriculture, Livestock, and Supply
MDA	Ministry of Agrarian Development
MASTER	Movement of Landless Farmers
MPA	Movement of Small Farmers
MEC	Ministry of Education and Culture
MIRAD	Ministry of Agrarian Reform and Development
MMC	Movement of Peasant Women
MST	Movement of Rural Landless Workers
NB	Grassroots
NGO (ONG)	Non-governmental Organization
PAA	Program for the Acquisition of Food
PCB	Brazilian Communist Party
PENSA	Program for Studies of the Agro-Industrial Business System
PNAE	National School Meals Program
PNAPO	National Policy for Agroecology and Organic Production
PNRA	National Agrarian Reform Plan
PR	Paraná
PROCERA	Special Credit Program for Agrarian Reform
PROMET	Methodological Project
PRONAF	National Program for Strengthening Family Agriculture
PRONERA	National Education Program for Agrarian Reform

PTA	Project Alternative Technology
RAP	Popular Agrarian Reform
RS	Rio Grande do Sul
SACA	Seminar for the Evaluation of Agroecology Courses
SCA	Settlers' Cooperative System
SESI	Industry Social Service
SPCMA	Sector for Production, Cooperation, and Environment
SIBRATER	Brazilian System for Technical Assistance and Rural Extension
SNA	National Society for Agriculture
SNCR	National System for Rural Credit
SSR	Rural Social Services
SUPRA	Agrarian Reform Oversight
TC	Community Time
TE	School Time
TS	Social Technology
UDR	Ruralist Democratic Union
ÚNICA	National Union for the Sugarcane Industry
ULTAB	Union of Farmer and Farm Workers
UEM	State University of Maringá
UFPR	Federal University of Paraná
UnB	University of Brasilia
UNESCO	United National Educational, Scientific, and Cultural Organization
UNICEF	United Nations Children's Fund
UNILA	Federal University of Latin America Integration
UNIOESTE	State University of West Paraná
USAID	United State Agency for International DevelopmentIntroduction

Introduction

Interest in the debate on "sustainable development", ecosocialism, agroecology and healthy food production is growing in Europe and around the world. Likewise, interest in getting to know more about anti-systemic struggles in Brazil and other Latin American countries is flourishing, especially concerning the actions and stances that anti-capitalist social movements have been developing regarding the production and reproduction of life.

In a previous book, *The World of Associated Labour and Embryos of an Education Beyond Capital,* one of the authors tried to address the multiplication of associated labour, communal property, and an education beyond capital. In the book you hold in your hands, we try to shed light on the "environmental" struggles of the Landless Worker's Movement (MST, an acronym for the Portuguese *Movimento dos Trabalhadores Rurais Sem Terra*) that foreshadow a new way of relating to nature and the educational struggles that point towards what Mészáros calls an education beyond capital.

The Landless Worker's Movement is one of the most important social movements of today because it combines land rights struggles with environmental struggles, international struggles with national struggles, articulating immediate struggles with a larger struggle for another kind of society, the class struggle with the gender struggle. Its defence of agroecology differs from ecocapitalist proposals or proposals orbiting around "sustainable development". We hope, then, to show the readers from outside Brazil some of this struggle's dimensions.

∴

We have watched in terror as social and environmental crimes multiply in Brazil. To recall just a few facts from the last decades: the murders of Chico Mendes and Doroty Stang, the massacres at Corumbiara[1] and Eldorado dos

---

[1] The Corumbiara massacre was the result of a violent conflict that took place on the 9th of August 1995 in the municipality of Corumbiara, in the state of Rondônia. The conflict began when police officers clashed with landless people who were occupying an area, resulting in the deaths of 100 people, according to the social movements, and 23 people (7 of whom are missing), according to the police. In August 1995, around 600 peasants had mobilised to take over the Santa Elina farm. In the early hours of the morning of the 9th, armed gunmen, recruited from large farms in the region, as well as military police soldiers with their faces covered, began attacking the camp.

Carajás,[2] the murders of leaders of the Peasant Leagues from the Brazilian Communist Party (PCB) and from Northeast Brazil in the 1960s, the escalating rate of murders of indigenous people, Quilombolas,[3] landless workers and squatters, the crimes committed by mining companies in Bento Gonçalves and Brumadinho, oil spills in the Northeast, and planned forest fires in the Amazon.

With its mind-manipulation techniques, Capital barrages us with the latest celebrity gossip, making us quickly forget the general meaning of these humanitarian and environmental crimes. It also leads us to believe that socio-environmental collapse must be solved within the framework of capitalist society, "through the consumption habits of individuals", without questioning the enormous power of transnational corporations and the State to destroy the conditions for life on earth and in promoting programmed obsolescence.

On the social level, right-wing forces had a historic opportunity with the 2016 coup that ousted President Dilma, which was followed by an avalanche of reforms to destroy social rights: the so-called "end of the world" amendment,[4] the labour and pension reforms, and laws furthering the commodification of education and health. The 2016 coup also enabled the rise of Jair Bolsonaro to power (2019–2022) through a fraudulent election plagued by widespread digital manipulation.

From a historical perspective, the Brazilian countryside has undergone a large restructuring process since the 1960s. The military regime (1964–1985) called the destructive advance of capital the "new agricultural frontier". In the case of the Amazon, the motto was "Integrate [the Amazon into Brazil] so we don't give it away". In the book "The dictatorship of big capital" (1981), Octavio Ianni recounts the destructive advance of capital towards new regions and borders. He also presents us with the emergence of new corporations in the south and southeast (pork, chicken, and ox slaughterhouses), as well as the installation of large transnational corporations producing pesticides, synthetic

---

2 It was Wednesday, April 17th, 1996, at around 4pm. Around 1,500 people were camped out in Eldorado do Carajás, in the south-east of Pará, in protest. The aim was to march to the capital Belém and get the Macaxeira farm, occupied by 3,500 landless families, expropriated. The march that had begun on 10 April was brought to a bloody halt in an attack by the Military Police that became known worldwide as the Eldorado do Carajás Massacre. A total of 155 military police were involved in the operation that left 21 peasants dead, 19 at the scene of the attack, and two others who died in hospital.

3 Quilombolas (analogous to Maroons in North America and the Caribbean) are the descendants of formerly enslaved people who escaped and struggled for their land, where they remain to this day, living and working collectively.

4 Refers to Constitutional Amendment No. 95, which prohibits any increase in public spending above inflation for 20 years. The amendment was passed in 2016.

fertilizers, tractors, and agricultural implements, the pillars of the Green Revolution. New highways, ports, airports, and hydroelectric power plants were built aiming to set the proper conditions for big capital to produce. He also lays out the coexistence of "archaic" forms of labour in the countryside (in labour regimes that are analogous to slavery) and the emergence of a new "rural" proletariat.

Rural Brazil was restructured in a brutal fashion, based on the assassinations of leaders from the Peasant Leagues of the PCB (Brazilian Communist Party) and the Peasant Leagues of the Northeast, the closing of rural unions, theft of indigenous lands, assassination attempts, burning of crops, etc.

The military had some colonization policies, but they obviously did not carry out agrarian reform. In the "re-democratization" period – which Florestan Fernandes called the "institutionalization of the dictatorship" -, some policies for the creation of landless worker settlements emerged during the Collor (1990–1992), Fernando Henrique Cardoso (1995–2002), and Lula governments, largely because of pressure by social movements in the countryside and international organizations.

Then we come to the Bolsonaro administration. From a social and environmental point of view the Bolsonaro campaign nodded emphatically to agribusiness with an agenda calling for hardened repression of rural social movements and total freedom for capital to advance towards new "virgin" areas. Bolsonaro even said that the "Indians would not get a single square inch of land" and compared Quilombolas to oxen, by measuring their weight in arrobas.[5] At the 2018 edition of the agribusiness event Agrishow, figureheads and politicians connected to this industry, locally known as *Ruralistas*, said they did not feel Geraldo Alckmin, a former governor of the state of São Paulo who was running for president at that point in time was a steady ally, as he hesitated in acquiescing to some agribusiness demands, and so they quickly migrated their support to Bolsonaro.

In January 2019, the crime at Brumadinho occurred.[6] Over 250 people died, and once again we had an ecosystem in shambles. As if the astounding crime

---

5 T.N.: The Arroba is a custom unit of weight, mass or volume, used in Spain, Portugal and their former colonies. In Brazil, it is almost exclusively used to measure the weight of livestock, thus the offensive nature of referring to the weight of humans with this unit.
6 T.N.: The rupturing of the Brumadinho Dam took place on 25 January 2019, when a tailings dam holding refuse from an iron mine burst, releasing a massive mudflow that flooded the city of Brumadinho, in Minas Gerais, leading to 270 deaths and the destruction of the town's infrastructure.

at Bento Gonçalves-Mariana in 2015[7] was not enough, Brumadinho followed a few years later. Shortly after the crime, an official declared that "the entire state of Minas Gerais should be in a state of alert".

In November 2018, the State Environment Chamber of Minas Gerais voted to reopen the Brumadinho dam, which had been deactivated for three years. Few seemed to remember the tragedy at Mariana, which had taken place on November 5, 2015. And, in a vote of 7 for and 1 against, the Chamber permitted the dam to operate. On that occasion, the only "civil society" representative, who voted against the measure, declared her vote as follows: "This borders on insanity". For us, it was insanity itself. Also in the state of Minas Gerais, at Mineirão Stadium, Germany eliminated Brazil from the 2014 FIFA Men's World Cup with the same score of 7x1. It was traumatic, but we dealt with it. The 7x1 score that allowed the dam to operate announced a true crime, different from the football tragedy, since it brought such harmful consequences for Brazilian people and ecosystems.

Intellectuals such as David Harvey and Jean Ziegler show us that Brazil has become one of the biggest stages for a new phase of "primitive" accumulation, based on the occupations of new lands in "virgin" regions by capital. According to the last IBGE agricultural census (2019), Brazil has undergone another land concentration process over the last 10 years. Not only that, but it revealed over one million unemployed in the countryside and the sale of one million tractors. Chronic problems in Brazil, such as access to land by peasants, hunger, malnutrition, and commodity exportation, are perpetuated in an increasingly dramatic fashion (Ziegler, 2013; Castro, 1980).

The reports of the Pastoral Land Commission (CPT) reveal an escalation of murders, including during the PT governments. As capital advances to new agricultural frontiers, such as southern Pará, southern Maranhão, Tocantins, and western Bahia, murder rates and leadership assassination attempts have increased in these areas.

In this scenario we must highlight the direct and indirect destruction of the National Education Program in Agrarian Reform (Pronera), a hard-fought achievement of the social movements of the countryside during the Fernando Henrique Cardoso (1995–2002) government. Pronera has contributed to the eradication of illiteracy in rural areas, created technical courses in agroecology, higher education courses in pedagogy, history, geography, veterinary science, agronomy, as well as graduate specialization courses and even master's

---

7   T.N.: The rupturing of the Mariana Dam took place in November 5th, 2015, leading to 19 deaths and insurmountable environmental damage to major river basins in the region.

degrees. This action, in tandem with others that destroy our fragile public educational system, once again signals that Brazil's wealthy classes only care about: a) the maintenance of the population in absolute ignorance, with the multiplication of illiteracy and functional illiteracy; b) the complete commodification of education; and c) a few training policies for the rare regions where there is industry or some demand for the education of the workforce.

The barbarism promoted by financial capital has brought harmful consequences for the working class around the world. The looting of public funds, expropriation of houses (like in the 2008 crisis), partial or complete destruction of the welfare state in Europe and the few constitutional rights won in the "unwelfare state" in Latin America are but a few examples. We see increases in the cost of living of the working class, the end of dignified retirement, and the destruction of educational and public health systems. In other words, the destruction of the conditions of social reproduction under capitalism steadily advances.

The voracity of globalized capital, with its "virtual senate" that decides the allocation of capital, does not respect popular decisions, tramples parliaments, and promotes coups all over the world. To mention only examples in Latin America, in recent years we have seen political arrests of presidents, irregular impeachments and more recently massacres in popular rebellions in Ecuador, Chile, Bolivia, Honduras, and Haiti, demonstrating the cruelty of the property-hoarding classes.

In addition to demanding from parliaments the total freedom to reproduce itself, destroying rights that were hard won by the working class, fictitious capital[8] also promotes a wide ideological manipulation and the stimulation of fascist processes through techniques including hybrid warfare and technological terrorism.

The destructive production of large transnational corporations (banks, insurance companies, mining companies, contractors, automobiles, the military complex, etc.), based on the expanded reproduction of capital and the programmed obsolescence of goods, generates large-scale socio-environmental crimes, as we have seen above, creates unbearable cities, and steals land and other resources that are strategic in the new world geopolitics. Furthermore,

---

8 Fictious capital: In common sense, the concept of financial capital is more commonly used, but it's more accurate to call it fictitious capital. It's not just banks, but also insurance companies, billionaire investors who have partially been able to make the reproduction of capital independent from the material bases of labour exploitation. See, for example, the book by Francois Chesnais (1995), *La Mondialisation du capital*.

the great powers generate medium- and low intensity wars that kill on an unprecedented scale with no shame.

In the world of labour, Taylorist-Fordist forms[9] are combined with the forms of the flexible accumulation regime[10] and, more recently, uberization and other forms of work analogous to slavery. In this context, overexploitation of labour, underemployment, and mass unemployment become part of the dramatic reality of nations the world over.

This book depicts the rural worker's resistance to the advancement of destructive production. It also socializes the results of research that shows us the foreshadowing of alternative forms of work based on agroecology, cooperation, and cooperativism, as well as the emergence of agroecology schools from one of today's most important social movements, the Landless Workers' Movement.

Arisen from the entrails of the socio-metabolism of capital, these novel forms of production and life have enormous emancipatory potential. They can move forward, but they can also quickly fizzle out if workers around the world don't get off the defensive. Mészáros (2002) believes that we must move not only beyond the neoliberal model, but beyond capital. Social movements are not only contesting neoliberal capitalism, but creating practical alternatives that can lead the way in the transformation to another type of society.

The MST's struggles to practice agroecology show us, in theory and in practice, the potential of food sovereignty, popular agrarian reform, gender equality in countryside, associated work, healthy food production and of an education beyond capital. They distance themselves from eco-capitalism, which tends to ignore the agrarian issue and stimulate actions in "corporate social responsibility", as well as from capitalist cooperativism, which moves in function of the expanded reproduction of capital.

These actions give rise to a criticism of land tenure and use in Brazil – wrought in iron and fire by large landowners – to the overexploitation of labour and the production of commodities for the foreign market. This cycle of commodity

---

9 Taylorist-Fordist forms: forms of work based on specialised and repetitive work, with rigid machines and generally unions tied to the state.
10 Flexible accumulation regime: According to David Harvey, flexible accumulation is based on the disruption of labour markets, temporary contracts, the incorporation of immigrant labour, etc. It manifests itself differently in different aspects: in manufacturing processes through the transposition of factories to other countries or regions; in the production of goods through just-in-time processes, batch orders, etc.; in financial markets through deregulated transactions relating to foreign exchange, credit and investments.

production and capital appreciation generates hunger and malnutrition in a country with a vast wealth of land and sun.

It is also worth remembering that four centuries of land concentration (not only command the dependent and associated economic insertion) of our bourgeoisie but also politically command our economic subsystem in the globalized capitalist system (Ianni, 1981). This political command prevents or hinders the struggles of social movements for land, education, health, decent work, etc.

The industrialization of agriculture, besides creating a vast business for financial capital, places the State at its service to create the general conditions for the production and reproduction of agribusiness, giving rise to a large market for pesticides, synthetic fertilizers, tractors, agricultural implements and transgenic seeds, the foundations of the "Green Revolution", and stimulating monoculture on large properties. Furthermore, it subordinates peasants, who are swallowed up by the siren song of the "Green Revolution", and end up in debt, working for a bank.

Our research shows that agroecology schools are the result of a practical opposition of the MST to the advance of agribusiness and the Green Revolution. They aim to train agroecological technicians to carry out the agroecological transition in the settlements. They cultivate the scientific principles of agroecology, are theoretically opposed to the advance of the Green Revolution, are based on the self-organization of students (experience of direct democracy in schools), in educational moments (study time, leisure time, working time, commissions time, etc.) and in the relationship between school time and community time. These schools, in terms of content, form and objectives, differ greatly from state schools and even more from private schools.

At the same time, however, contradictions emerge in the productive and educational experiences of the MST. Part of the practices of cooperation, cooperativism and agroecological production have gained an economic bias, solely and only created to have access to public funds or for immediate survival. Regarding agroecology schools, part of the trained technicians cannot get a job in the settlements or even a job in the area. The principles of self-organization, relationships between study and work, school and settlements, and school time and community time, among others, are all partially distorted or run into some difficulty.

1 The Book's Organization

Chapter 1 of this book is called "The current state of "primitive accumulation"". In it we seek to recover the debate on land theft and fencing, trying to show that this process has not been interrupted. After discussing the violent processes that separated workers from the means of production and led to the formation of the working class as a dispossessed class, we highlight the neocolonialism of the nineteenth century and demonstrate the new cycle of theft and land enclosure, especially from the 1960s onwards. Researchers Rogério Fernandes Macedo and Fábio Castro collaborated in this chapter.

In "Destructive production and agroecology", Chapter 2 we try to show the destructive advance of transnational corporations and to systematize the work of the agroecology's main theorists, namely: Miguel Altieri, Sevilla Guzman and Pinheiro Machado. Chapter 3, entitled "Sustainable Development, Agroecology and Ecosocialism", offers some of the contributions of sustainable development theories, but mainly their limits. We then point out the contributions of ecosocialism to the conception of socialist agroecology.

In Chapter 4, "Green Revolution" in Brazil, rural extension and the fight to establish MST Agroecology, we analyse that agroecology as a socioproductive model, became more consistent on the Movement of Landless Rural Workers' agenda (MST) during the development of its fourth National Convention held in the year 2000 in Brasilia. The convention's formation was characterized by its intense debate proceedings and reflections regarding the development of a grassroots agricultural project, as well as a critique of the Green Revolution model and of its opposition to agribusiness. The critique of the Green Revolution and the opposition to agribusiness has been gaining notoriety within MST since the preparation and completion of its III National Convention held in 1995. Since its III Convention, MST began thinking over the fact that the conventional model of agriculture which establishes the Green Revolution as a foundation imposes a series of negative and contradictory consequences to agrarian reform. However, it was only in its IV Convention that MST shifted gears making a commitment to develop an alternative and grassroots program for the base, rooted in another technological model – agroecology.

Chapter 5 is called "Perspectives and Dimensions of Agroecology". Agroecology is the main epistemological and practical principal which is still under construction within MST's territories and by Brazilian researchers. Although it is still an open question and under discussion, it is noticeable that workers that comprise MST are united in their understanding of the agroecological framework, a common means of rejecting the framework of the *Green Revolution*. The agroecological concepts and practices aren't homogeneous, and despite MST having agroecology as a resistance and alternative to agribusiness, it is still a disputed project, which isn't exempt from the abuse of agribusiness. The objective of this chapter is to present the main schools of thought that contributed and contribute to its historical construction from the agroecological perspective, taking into consideration the conceptual polysemy that arises from the different adopted epistemological branches, be they more reformists or more radical and revolutionary.

In Chapter 6, "Transnational corporations, and the MST's agroecological agenda and agroecology schools", we continue this debate and try to show how and why the agroecological agenda became a part of the MST and the reasons that led them to create technical agroecology schools. Chapter 7, "The political economy of the "Green Revolution", Agroecology and the MST Agroecology Schools", is co-authored where we criticize the so-called "Green Revolution", showing attempts to build an alternative production proposal, so-called agroecology, as well as the wealth and contradictions of the MST's schools.

Chapter 8, "Cooperation and workers' cooperatives of MST". We note that the State of São Paulo provides a privileged standpoint to observe the restructuring of the countryside and the (im)possibility of an agrarian reform that aims to create the general conditions for the development of cooperation, cooperativism, and agroecological agrarian reform settlements. The first half of the chapter makes a brief historical overview of the political economy of the countryside in the State of São Paulo, giving special emphasis to São Paulo's autocracy. The second part shows the limits and contradictions of cooperation and workers' cooperatives in the MST settlements in the face of the new agribusiness offensive in the State.

In Chapter 9 "Fundamentals of the Emancipated School of Labour: the contribution of Soviet pedagogues" we seek to show fundamentals of the school of labour. We realized that, directly and indirectly, social movements drink in the theory of Soviet pedagogy, which had as principles – at least in the first phase of the revolution – the construction of a society that was not based on the exploitation of work, the self-organization of schools and the educational system, using thematic complexes as a way to combat positivism and connect theory with practice and inserting schools in the struggles of their historical period.

∙∙∙

Our book is a presentation of the results of three studies, two of them funded by FAPESP (Process 2014/19013–8), (Process 2020/01666–6 in progress) and the other by CNPq (Process No. 473180/2014–6). In conclusion, this book is the result of politically engaged research. In Latin America, since the Córdoba Reform of 1918, a trend has emerged in the educational field, of teachers who are concerned with recovering the unity between theory and practice, between critical reflection and radical action. As far as possible, we have helped the social movements of the countryside and the city to materialize and theorize agroecology and to practice the principles of self-organization and emancipation of work in agroecology schools.

CHAPTER 1

The Current State of "Primitive Accumulation": Land Theft and Enclosures in the Twentieth and Twenty-First Centuries

1   Introduction

> So-called primitive accumulation, therefore, is nothing but the historical process of divorcing the producer from the means of production. It appears as primitive, because it forms the pre-historic stage of capital and of the mode of production corresponding with it.
> KARL MARX, 2012

> The single key factor in the revival of debate on enclosures is undoubtedly neoliberal globalisation itself. A huge round of dispossession and accumulation is currently under way involving a global assault on customary use rights; the transformation of common resources into private property; and the introduction of market mechanisms into all aspects of social life. For instance, the current transfer of land, water and forests into private hands in India is, as Arundhati Roy suggests, 'a process of barbaric dispossession on a scale that has no parallel in history'. The British Marxist historians paid a great deal of attention to the enclosure of the commons and the current round of enclosure gives their work a new urgency. The British case does not offer a 'one-size fits all' model for contemporary capitalism, but this work does provide substantial intellectual tools for thinking about the latest round of global enclosures.
> EDWARDS, 2017

The quote from Steve Edwards (2017) above, and the works of David Harvey (1993; 2012) and Arundhati Roy (2001) suggest that the globalisation of capital, from the 1960s to now, boosted a new cycle of dispossession and land enclosure that gives researchers around the world "a new urgency". One of the most important chapters of Karl Marx's book "Capital" (1985) is called "So-called primitive accumulation". In it, Marx recounts the enclosure of common lands in Britain. With historical cunning, he presents us to the enclosure of communal lands that gave rise to the dispossessed proletariat that would flood English cities in the role of factory proletarians, forming an industrial reserve army.

Less known are Marx's (2012) writings on Russia, which in the Boitempo Edition were called "The Class Struggle in Russia". In them, Marx shows us the persistence of the rural communes in Russia and their advantage in a possible and necessary transition to communism. Against those who advocated the complete extinction of the communes in the name of "progress", Marx advocates a dialectical preservation and updating of these experiences of collective life, collective decisions, and collective property. It also warns that the separation between the producer and the means of production only took place in a radical way in England, even if the other countries of Western Europe went through the same process [*mouvement*] (Marx, 2012, p. 114).

In the text "Value, price and profit" (Marx, 2012, p. 111) he notes that what economists call primitive or original accumulation should be called *original expropriation* and follows:

> We should find that this so-called *original accumulation* means nothing but a series of historical processes, resulting in a *decomposition* of the *original union* existing between the labouring Man and his Instruments of Labour. Such an inquiry, however, lies beyond the pale of my present subject. The *separation* between the Man of Labour and the Instruments of Labour once established, such a state of things will maintain itself and reproduce itself upon a constantly increasing scale, until a new and fundamental revolution in the mode of production should again overturn it, and restore the original union in a new historical form.
>
> MARX, 2012, p. 111 – italics in the original

This chapter aims to recover the concept of "primitive accumulation" and show how current it is. Authors such as David Harvey (2004) and Walter Porto Gonçalves et. al. (2016) have argued that there was no "big bang" of land enclosure. If they are right, incessant land theft was present in the twentieth century, and was leveraged with the globalization of capital from the 1960s onwards.

As we will see, Brazil is the stage for one of the current chapters of the new phase of "primitive accumulation": we are one of the countries with the most land theft in the twentieth century, we are champions in the murder of indigenous people, *Quilombolas*,[1] squatters, and landless people, ahead of countries

---

[1] Quilombolas (analogous to Maroons in North America and the Caribbean) are the descendants of formerly enslaved people who escaped and struggled for their land, where they remain to this day, living and working collectively.

such as Colombia and Indonesia. And, it seems, Bolsonaro's government is providing new impetus to land grabbing in the country.

2 So-called Primitive Accumulation

In the chapter entitled "So-called primitive accumulation", Marx (1985) addresses some historical structures and dynamics that ended up shaping the classical social-capital relationship, articulated with the exploitation of labour, capitalist production, added value, and the accumulation of capital. He undertakes an effort to understand how original or primitive accumulation took place, preceding capitalist accumulation, an accumulation which is not the result of the capitalist mode of production but its point of departure". (Marx, 1985, p.261). With this in mind, he warns readers:

> The capital-relation presupposes a complete separation between the workers and the ownership of the conditions for the realization of their labour. As soon as capitalist production stands on its own feet, it not only maintains this separation, but reproduces it on a constantly extending scale. The process, therefore, which creates the capital-relation can be nothing other than the process which divorces the worker from the ownership of the conditions of his own labour; it is a process which operates two transformations, whereby the social means of subsistence and production are turned into capital, and the immediate producers are turned into wage-labourers.
>
> MARX, 1985, p.262

So, one can identify a process that restructures the ownership of the conditions of the labour of direct producers, under which their labour forces and their social means of subsistence and production have been converted into capital and integrated into the historical process of primitive accumulation at the dawn of the capitalist era. In summary, Marx records:

> So-called primitive accumulation, therefore, is nothing else than the historical process of divorcing the producer from the means of production. It appears as 'primitive' because it forms the pre-history of capital, and of the mode of production corresponding to capital.
>
> MARX, 1985, p.262

The author (Marx, 1985) situates this process in the lower Middle Ages: despite capitalist production first appearing sporadically "as early as the fourteenth or fifteenth centuries in certain towns of the Mediterranean, the capitalist era dates from the sixteenth century " (Marx, 1985, p.263), a time of broad decline of sovereign cities and servile feudal relations. Born out of a series of contradictory historical dynamics, primitive accumulation and the social-capital relationship will, in their early days, play a double role, tinted with positivity and liberation and tainted with negativity and subjection. Marx (1985) explains the terms of this contradiction as follows:

> To become a free seller of labour-power, who carries his commodity wherever he can find a market for it, he must further have escaped from the regime of the guilds, their rules for apprentices and journeymen, and their restrictive labour regulations. Hence the historical movement which changes the producers into wage-labourers appears, on the one hand, as their emancipation from serfdom and from the fetters of the guilds, and it is this aspect of the movement which alone exists for our bourgeois historians. But, on the other hand, these newly freed men became sellers of themselves only after they had been robbed of all their own means of production, and all the guarantees of existence afforded by the old feudal arrangements.
> MARX, 1985, p.262

Within this process, capitalists were located (Marx, 1985) at the spearhead of the theoretical and practical transformations at play, providing (for the benefit of their particular interests) the transient emancipation of society in general, of feudal structures and dynamics, controlled by the masters, the church, and craft corporations. In this perspective, the rise of the capitalists:

> appears as the fruit of a victorious struggle both against feudal power and its disgusting prerogatives, and against the guilds, and the fetters by which the latter restricted the free development of production and the free exploitation of man by man. The knights of industry, however, only succeeded in supplanting the knights of the sword by making use of events in which they had played no part whatsoever.
> MARX, 1985, p.262

Through these events, primitive accumulation will occur, necessary as it is in the formation of the social-capital relationship. Both construct and influence each other simultaneously, implying a process of significant emancipation, which will create positive waves of impact for direct producers within a relatively short historical horizon. That is to say: the emancipatory essence of the revolutionary capitalist processes should be constrained and restricted, for the sake of the emergence and, especially, the consolidation of the social-capital relationship; the aspirations of emancipation of those direct producers needed to be contained, under penalty of exceeding certain historical limits, after which the nascent and current system could no longer be called capitalism. In short: the capitalist revolution is followed by its setback, namely its counter-revolution.

Within it, the profound transformation of land tenure, which structured relations of feudal production and reproduction, was a decisive episode. In his attempt to understand this phenomenon, Marx (1985) goes on to explore the English experience. With a relative richness of detail, the author demonstrates a long process of expropriation of feudal land ownership, making use of the examination of legislation (first, protective; then, destructive of workers) established by successive reigns, notably with regard to the "turning of arable into pasture land, begins at the end of the fifteenth century and extends into the sixteenth". (Marx, 1985, p. 268). The author will say:

> The prelude to the revolution that laid the foundation of the capitalist mode of production was played out in the last third of the fifteenth century and the first few decades of the sixteenth. A mass of 'free' and unattached proletarians was hurled onto the labour-market by the dissolution of the bands of feudal retainers, who, as Sir James Steuart correctly remarked, 'everywhere uselessly filled house and castle'.* Although the royal power, itself a product of bourgeois development, forcibly hastened the dissolution of these bands of retainers in its striving for absolute sovereignty, it was by no means the sole cause of it. It was rather that the great feudal lords, in their defiant opposition to the king and Parliament, created an incomparably larger proletariat by forcibly driving the peasantry from the land, to which the latter had the same feudal title as the lords themselves, and by usurpation of the common lands. The rapid expansion of wool manufacture in Flanders and the corresponding rise in the price of wool in England provided the direct impulse for these evictions.
>
> MARX, 1985, p. 264

It is noteworthy that the revolution of the Flemish productive forces promoted economic stimuli, which triggered several changes in the midst of primitive accumulation: 1) it changed the relationship of the feudal lords with servile labour, which would increasingly be displaced from the countryside in favour of the extraction and direct appropriation of land income; 2) it changed the relationship of the feudal lord with the land, which was now accumulated to be used as pasture; 3) it modified the relationship of feudal lords with money, since the new generations began to see it as the foundation of their socioeconomic and political power; 4) it restructured social relations at the top rungs of society, even leading to the emergence of a distinction of identity between the old and the new feudal nobility, as well as between their respective bases of economic support: the old feudal nobility "ad been devoured by the great feudal wars. The new nobility was the child of its time, for which money was the power of all powers. Transformation of arable land into sheep-walks was therefore its slogan". (Marx, 1985, p. 264).

Marx (1985) demonstrates that this expropriation of land affected direct producers, the feudal state, the Catholic Church and, finally, the communal property of land, which "was an old Teutonic institution which lived on under the cover of feudalism". (Marx, 1985, p. 268). Faced with all the extreme poverty and the state of calamity created by this violent process of expropriation and restructuring of feudal land ownership, several laws were enacted to impose some control over such dramatic consequences. For 150 years, such efforts at legal control proved to be useless and there were notable setbacks throughout the eighteenth century, through which "he law itself now becomes the instrument by which the people's land is stolen, although the big farmers made use of their little independent methods as well". (Marx, 1985, p. 269). They were the *Enclosure Bills* namely, "decrees by which the landowners grant themselves the people's land as private property, decrees of expropriation of the people". (Marx, 1985, p. 269). As a quantitative reference (Marx, 1985), between 1810 and 1831, that is, in only 21 years, no less than 3,511,770 hectares (or 1,404,708 hectares) of communal land were stolen from direct producers and transferred to landlords, promoting dramatic population displacement to industrial cities, where producers were profusely converted into workers, freed from feudal ties and imprisoned to capitalist ones.

After the Enclosure Bills, the last land expropriation tactic was called the:

> 'clearing of estates', i.e. the sweeping of human beings off them. All the English methods hitherto considered culminated in 'clearing'. As we saw in the description of modern conditions given in a previous chapter, when there are no more independent peasants to get rid of, the 'clearing' of cottages begins; so that the agricultural labourers no longer find on the soil they cultivate even the necessary space for their own housing.
>
> MARX, 1985, p. 271

According to the author, land clearing was widely and exemplarily used in Upper Scotland. There, several clans dominated the land structure and at the head of each of them was a clan leader, also called "great man". They faced each other in intense wars, which, over time, were contained by the English government. From then on, the "great men", by their own authority, "they transformed their nominal right to the land into a right of private property, and as this came up against resistance on the part of their clansmen, they resolved to drive them out openly and by force". (Marx, 1985, p. 272). Such events date from approximately 1745/46 and these expelled Gaelic peasants were prohibited from emigrating (Marx, 1985), leaving them to thicken the mass of workers in manufacturing cities such as Glasgow. Such a method of clearing became the rule, including during the nineteenth century and, to exemplify this fact, Marx (1985) refers to the clearing carried out by the Duchess of Sutherland, as follows:

> This person, who had been well instructed in economics, resolved, when she succeeded to the headship of the clan, to undertake a radical economic cure, and to turn the whole county of Sutherland, the population of which had already been reduced to 15,000 by similar processes, into a sheep-walk. Between 1814 and 1820 these 15,000 inhabitants, about 3,000 families, were systematically hunted and rooted out. All their villages were destroyed and burnt, all their fields turned into pasturage. British soldiers enforced this mass of evictions, and came to blows with the inhabitants. One old woman was burnt to death in the flames of the hut she refused to leave. It was in this manner that this fine lady appropriated 794,000 acres of land which had belonged to the clan from time immemorial. She assigned to the expelled inhabitants some 6,000 acres on the sea-shore – 2 acres per family. The 6,000 acres had until this time lain waste, and brought in no income to their owners. The Duchess, in the nobility of her heart, actually went so far as to let these waste lands at an average rent of 2s. 6d. per acre to the clansmen, who for centuries had shed their blood for her family. She divided the whole of the stolen land of the clan into twenty-nine huge sheep farms, each inhabited by a single family, for the most part imported English farm-servants. By 1825 the 15,000 Gaels had already been replaced by 131,000 sheep.
> MARX, 1985, p. 273

The disruption of the feudal land base and the rural exodus corresponded to a part of the martyrdom to which expropriated producers were subjected. Later, in industrial cities, this drama would expand and increase. Despite the extreme dependence of the bourgeois elites on such workers, their capacity to absorb such a supply of labour power was limited by the concrete conditions of production. As a result, legions of poor beggars in subhuman conditions have emerged. Although functional for the strategic lowering of wages, these elites contradictorily imposed the extermination of these populations, subjecting them to strict rules of conduct, impossible to be followed: Marx (1985) calls them Bloody Legislation. Let's take a look at some examples of English law:

> Edward VI: A statute of the first year of his reign, 1547, ordains that if anyone refuses to work, he shall be condemned as a slave to the person who has denounced him as an idler. (...) If the slave is absent for a fortnight, he is condemned to slavery for life and is to be branded on forehead or back with the letter S; if he runs away three times, he is to be executed as a felon. The master can sell him, bequeath him, let him out on hire as a slave, just as he can any other personal chattel or cattle. (...) Elizabeth, 1572: Unlicensed beggars above 14 years of age are to be severely flogged and branded on the left ear unless someone will take them into service for two years; in case of a repetition of the offence, if they are over 18, they are to be executed, unless someone will take them into service for two years; but for the third offence they are to be executed without mercy as felons.
>
> MARX, 1985, p. 275–6

In this way, farmers, expelled from their land base and converted "into vagabonds, and then whipped, branded and tortured by grotesquely terroristic laws into accepting the discipline necessary for the system of wage-labour". (Marx, 1985, p.277). This is the weight of bloodthirsty legislation on those who, by socioeconomic and political determinations of the nascent system of capital production, could not sell their labour-power. On the other hand, a certain contingent ended up succeeding in the vital attempt to sell its labour-power: these were also subjugated by inhuman laws, dedicated to regulating their wages. According to the author (Marx, 1985), they originated in England between 1349 and 1813, very long term, therefore. Let us see how Marx (1985) demonstrates the content of this legislation, referring to the *Statute of Labourers*, of 1349. He said:

> A tariff of wages was fixed by law for town and country, for piece-work and day-work. The agricultural labourers were to hire themselves out by the year, the urban workers were to do so 'on the open market'. It was forbidden, on pain of imprisonment, to pay higher wages than those fixed by the statute, but the taking of higher wages was more severely punished

> than the giving of them (similarly, in Sections 18 and 19 of Elizabeth's Statute of Apprentices, ten days' imprisonment is decreed for the person who pays the higher wages, but twenty-one days for the person who receives those wages).
>
> MARX, 1985, p. 278

After centuries of legislation devoted to lowering wages paid to workers, the capitalist mode of production, already in its manufacturing phase, held the economic and political means of wage compression, thus enabling the controlled abolition of these laws in 1813 (Marx, 1985). Of course, this abolition did lead to a right of association for workers, which was systematically fought by the bourgeoisie throughout the nineteenth century: only in the second half of that century "only against its will, and under the pressure of the masses, did the English Parliament give up the laws against strikes and trade unions (...)". (Marx, 1985, p. 279).

Furthermore, Marx (1985) deals briefly, in the chapter "So-called primitive accumulation", with the genesis of capitalist tenants. The author records that in England they have their origin in the bailiffs, but does not provide further details, saying only that in the second half of the fourteenth century, the bailiff is "replaced by a farmer, whom the landlord provides with seed, cattle and farm implements. The farmer's condition is not very different from that of the peasant", (Marx, 1985, p.280), distinguishing themselves only by exploiting more wage earners: such tenants would become sharecroppers. Marx (1985) notes that this form quickly gives way to the lessee itself, who "valorizes his own capital by employing wage-labourers, and pays a part of the surplus product, in money or in kind, to the landlord as ground rent.". (Marx, 1985, p.281). They will gain wealth throughout the sixteenth century in direct proportion to the impoverishment of rural workers (Marx, 1985). He explains:

> A further factor, of decisive importance, was added in the sixteenth century. At that time the contracts for farms ran for a long time, often for ninety-nine years. The progressive fall in the value of the precious metals, and therefore of money, brought golden fruit to the farmers. Apart from all the other circumstances discussed above, it lowered wages. A portion of the latter was now added to the profits of the farm. The continuous rise in the prices of corn, wool, meat, in short of all agricultural products, swelled the money capital of the farmer without any action on his part, while the ground rent he had to pay diminished, since it had been contracted for on the basis of the old money values.
>
> MARX, 1985, p.281

According to the author, the end of the sixteenth century witnessed the abrupt enrichment of lessees disproportionately in relation to *landlords*, stagnant, and inversely proportional to impoverished rural workers: it was "a class of capitalist farmers who were rich men in relation to the circumstances of the time". (Marx, 1985, p.281).

Marx's analysis is very distinct from those developed by representatives of political economy, such as Adam Smith and David Ricardo. It is also quite different from Max Weber's analysis in his classic book "The Protestant Ethic and the Spirit of Capitalism". We know Marx relied on classical political economy, especially the Labour Value Theory, but overcame it to the point of inaugurating the critique of political economy. For Marx, these thinkers "ignore" primitive accumulation, "hiding" the whole violent and bloodthirsty process of land grabbing in Europe that transformed producers into wage-earners. For Marx "the history of this expropriation is recorded in the annals of humanity with traces of blood and fire" (Marx, 2010, p. 786–87), therefore, very far from the docility of Max Weber's interpretation.

Criticizing Locke and Smith's view based on an alleged "original sin", Marx notes that for these:

> At a very remote time, there was, on the one hand, a laborious, intelligent and especially parsimonious elite and, on the other, a bunch of vagrants dissipating everything they had and even more [...] Thus, the former accumulated wealth and the latter ended up having nothing to sell, except their own skin. And from this original sin date the poverty of the large mass, which even today, despite all its work, still has nothing to sell but itself, and the wealth of the few, which grows continuously, although they have long ceased to work.
>
> MARX, 2010, p. 787

We saw in this chapter's opening that, for Marx, "so-called primitive accumulation" is nothing more than the historical process of the separation between producer and the means of production. It appears as "primitive" because it makes up the prehistory of capital and the mode of production that corresponds to it (Karl Marx, 2012). About the so-called "primitive accumulation" and the revolution in agriculture that led to the formation of the working class in England, Rosa Luxemburg says:

> In England, the new mode of production was introduced by a revolution in agriculture. The development of the wool industry in Flanders had provoked a great demand for it and prompted the English feudal

nobility to turn a large part of their arable land into sheep pastures, driving the peasants out of their lands and dominions. A large mass of workers who had nothing, proletarians, thus found themselves at the disposal of the first capitalist industry. The Reformation acted in the same direction, dragging with it the confiscation of the Church's goods, which were largely given, and others sold to abandonment, to the nobility and to the speculators, and where the peasant population was mostly driven away. Manufacturers and capitalist landowners thus found a poor, proletarianized population, fleeing feudal and corporate regulations and that, after the long martyrdom of a wandering life, the hard work in the Workhouses, the cruel persecutions of the law and the squabbles of the police, saw a port of salvation in wage slavery, at the service of the new class of exploiters. Then came the great technical revolutions that allowed the unqualified wage proletarian to be place aside or in place of the skilled craftsman.

LUXEMBOURG, 1977

Marx reminds us that the first bourgeoisie were merchants who accumulated capital to the point of being able to acquire means of production, hire workers, and exploit their labour-power by bringing them together in the same space (manufacturing) (Marx, 2010). From manufacturing to large industries, there is a qualitative technological leap in the exploitation of labour, increasingly based on relative value more than on more absolute value. In Marx's words, the "automatic machine system" inaugurated relative value, less and less based on physical exploitation and the extension of the working day or on the compression of work on the same working day. For him, the specific capitalist mode of production only arises with the advent of large industries.

3 The Enclosure and Theft of Land in Brazil in the 20th and 21st Centuries

One of the most engaging works of literature to understand "primitive accumulation" in the nineteenth century is "Heart of Darkness" by Joseph Conrad. He shows us the role of colonialism in Africa: in the name of "civilizing backward peoples", Europeans promoted one of the greatest waves of looting in the history of humanity, compared in proportion to colonialism on the American and Asian continent, driven by the great navigations of the 16th to 18th centuries.

Conrad indirectly shows us that what was at stake was an imperialist race for existing natural and strategic resources in Africa, in a context of disputes between monopoly capital supported by National States. Conrad very accurately narrates the point of view of Europeans, thirsting for the dominance of those lands and strategic resources and for the dominance of the world economy, deploying tricks to subjugate African peoples.

Likewise, João Bernardo (2004) describes a historical fact of the early twentieth century:

> An extreme case occurred in 1904 in Southwest Africa, when the German army employed such brutality to suppress the Hereros' revolt that between 75% and 80% of this population were slaughtered and between 23% and 18% were held in prison camps. All land and livestock were then confiscated, so there was nothing left for the survivors but to request employment in the service of the German settlers.

Arundhati Roy (2001), bringing this debate to the present day, notes that in India there is "a process of barbaric dispossession on a scale that has no parallel in history". Bernardo Mançano Fernandes (2013) also shows us that there a great cycle of land expropriation has been taking place in Africa over the last 40 years. Walter Porto Gonçalves (2017) reminds us that Brazil was one of the countries where processes of land enclosure were most prevalent in the twentieth century, especially from the corporate-military dictatorship (1964–85) on.

The history of Brazil and its role in the international division of labour since the 16th century are well known. The invasion of Brazil by the Portuguese is one child of the great Portuguese navigations, which aimed to conquer the best routes to the east. We were born for capitalism as a great exploitation colony, where everything that is produced is "exported" and everything that is needed is imported (the so-called colonial pact). Brazil is the birthplace of the huge sugar plantation, based on slave labour, without limits to the free exploitation of labour.

According to Carlos Cordovano Vieira (2019, p. 226), "Marx summarized in a polished sentence that 'the covert slavery of salaried workers in Europe needed slavery without disguises in the New World as a pedestal'". He quotes Marx once again:

> The discovery of the lands of gold and silver in America, the extermination, enslavement, and displacement of the native population to the mines, the beginning of the conquest and looting of the East Indies, the transformation of Africa into an outpost for commercial hunting of black-skinned bodies mark the dawn of the era of capitalist production. These idyllic processes are fundamental moments of primitive accumulation.

> Trade wars between European nations, with the world on the stage, then take place. (...) The distinct moments of primitive accumulation (...) in England at the end of the 17th century are systematically summarized in the colonial system, the public debt system, the modern tax system, and the protectionist system. These methods are based, in part, on the most brutal violence, such as in the colonial system. All, however, used the power of the State, the concentrated and organised violence of society, artificially activating a process that transforms the feudal mode of production into the capitalist mode, and to shorten the transition. Violence is the midwife of the old society, pregnant with a new one. It itself is an economic powerhouse.
>
> MARX, 1988, p. 275–276

The theft and enclosure of land in America from the sixteenth century on is certainly one of the most violent periods in human history. Pizarro, a Spanish man, can be considered one of the greatest killers in history. Along with his cronies, they exterminated about 5 million indigenous people in America. The history of colonial America is bathed in rivers of blood and killings "in the name of civilising barbarian peoples", in this case the indigenous peoples of the Americas. The history of the enclosure and theft of land in Brazil gains another chapter in the late nineteenth and early twentieth centuries, with the policies of colonisation and attraction of Italian, Spanish, German, Japanese, and Portuguese immigrants.

Our argument is that the practice of stealing and enclosing land, killing indigenous people, landless workers, and squatters and stealing their "territory" are not only part of the history of the sixteenth to eighteenth centuries. They are "natural" and incessant practices of the mode of production and reproduction of capital, like the "sunrise" of everyday capitalism.

In the 1950s-60s world capitalism undergoes a new restructuring. In Brazil there were numerous struggles for agrarian reform "by law or by strength". The 1964 coup begins a long process of land expropriation. The new phase of "primitive accumulation" in Brazil appeared under a pompous guise: the "new agricultural frontier". "Integrating so as not to give up" the Amazon, occupying these northern territories so as not to let the "gringos" steal it: these were the slogans of an alleged nationalist program for the Amazon. The construction of the Belém-Brasília highway, the Transamazônica highway, and the BR-364 highway, which connects Cuiabá to Porto Velho, could be highlighted here as important elements in this phase.

The agrarian policy of the corporate-military dictatorship created a true Brazilian Western and a new chapter in the history of the theft and enclosure of land in human history. Based on the "colonisation" policies in the regions of western Paraná, western Santa Catarina, Mato Grosso do Sul, Mato Grosso, Pará, Rondônia and southern Amazonas, many families of gauchos[2] headed north and spread to the west. Octavio Ianni (1981) called this period the "dictatorship of big capital".

The developments of this "new agricultural frontier" are now well known: indigenous people were trapped in reserves and continue to lose their territory to agribusiness. Landless squatters have lost their land. Social movements fighting for agrarian reform were strangled and their leaders exterminated. The "expansion of the agricultural frontier" brought with it the so-called "Green Revolution" package. Brazil has become one of the largest buyers of tractors and agricultural implements, synthetic fertilizers and pesticides, and more recently, of transgenic seeds. Many transnational corporations have settled here, thirsting for extraordinary profits. Banco do Brasil, a large state-owned bank, has created a variety of credit lines to favour this expansion. Roads, silos, ports, and airports were built to meet these demands. Technical, undergraduate, and postgraduate courses were created or restructured to meet the needs of this "new agricultural frontier". In Marx's terms, the "general conditions for the production and reproduction of capital" were created.

The ceaseless march of capitalism has gained new heights in the Fernando Henrique Cardoso (1995–2002) and Lula (2003–2010) governments. The main line of the Lula governments to rural areas, as we know, was the strengthening of agribusiness, including a new expansion of ethanol and sugar production plants. The expansion of soybean crops in the Central-West region pushed cattle raising on a large scale to the North. A new agricultural frontier called MATOPIBA emerged, expanding financial-agrarian capital towards the countryside of the states of Maranhão, Piauí, Tocantins, and Bahia. The consequences of this policy, focused on strengthening agribusiness, are also well known. Murders in rural areas have more than doubled over the last 10 years! (CPT, 2017).

Likewise, the crimes of Mariana and Brumadinho gained prominence in recent years and revealed the sordid bases of the Brazilian economy: an economy based on the exportation of primary products such as soybeans, ores, pork, chicken, and beef. An economy based on the exploitation of natural resources at all costs. Speaking of "whatever the cost", the last report of the Environmental Chamber of Minas Gerais on the authorisation or not of the

---

2   T.N.: Though currently the term "Gaucho" is used in Brazilian Portuguese to refer to any citizen native to the southernmost state of Rio Grande do Sul, originally it described the colonizers of the Pampas of southern Brazil, Uruguay and northeastern Argentina.

Feijão Dam in Brumadinho should be highlighted. The vote, as we all got to know, ended at 7–1.

In the 2014 World Cup Brazil was thrashed by Germany: we lost 7x1. The people were very sad, after all we lost at home and by a huge margin. But much more important than this football defeat was the 7–1 defeat in December 2018. The only person who voted against the permission of the dam – a representative of civil society – said that it "bordered on insanity". We saw a few months later that it didn't border on insanity. It was insanity itself. The dam collapse took lives and interrupted stories. "Profit over all, mud over all", someone wrote on the internet.

Karl Marx (2012) and Vladimir Lenin (2010) showed us that the capitalist state was shaped in the image and likeness of companies (in Marx's time), or trusts (in Lenin's time) primarily to meet the objectives of preserving private property, accumulating capital, and exploiting labour.

In the Brazilian case, mining laws were made by the mining companies themselves, by their lobbyists, and members of the executive and legislative. Many legislators from Minas Gerais, Goiás, Mato Grosso, and Pará are financed directly or indirectly by transnational corporations. Just a few weeks after the Samarco (BHP-Vale) disaster in Mariana, a law favouring corporations was passed. An ordinary citizen might ask: Now, how could it be? After an immense crime, a more favourable law? Representatives of big capital, who have their cake and eat it too in the construction or adaptation of the sector's legislation, are embedded throughout Minas Gerais's executive and legislative.

Oswaldo Sevá Filho (2019) argues that the territory's overwhelming spoliation took place in recent years due to the expansion of capitalist infrastructure. He highlights the energy, mining, and construction sectors as largely responsible for "clearing the ground", a term used by capital managers for what we are calling the new phase of "primitive accumulation". Plínio Sampaio Jr (2023) notes that Brazil has returned to the status of a large colony. Here, industry has less and less of a role. If this is true, ore extractions, water theft and privatisation, and exportation of primary products set the tone for this immense rural Brazil. The archaic becomes modern. Sugarcane, a symbol of colonial Brazil, becomes "the salvation of the countryside".

4   The Commodification of All Spheres of Life and Struggles of Resistance against "Primitive Accumulation"

In his book "Os Irredutíveis" (The Unyielding), Daniel Bensaid (2008) portrays the privatisation of services and social security, among others, and states that:

> the generalised privatisation of the planet extends to information, law (when private contracts prevail over the law of the land), solidarity (private insurance and pension funds as opposed to mutual insurance and social security), violence (in France, there are more private security guards and militias than "public" police), and even prisons.
>
> BENSAID, 2008, p. 65

He goes even further:

> As the land was already private property, now it is about the city's surroundings, with its protected neighbourhoods and its gated condominiums in urban conglomerates that favour public safety for the wealthy; water, with profitable distribution; air, with the project of a world market for the right to pollute; the frenzy of patenting, with the plausible prospect, from now on, of an embryo and clone market or private exploitation of the human genome. This narrowing of the public space is full of dangers to democracy, formal or otherwise.
>
> BENSAID, 2008, p. 65

He continues:

> Intimacy has also become a tradable commodity. The privatisation of the world has as its counterpart an increasing "publicization" of private life. This does not apply only to so-called "public" figures, who display their private image due to the need for media promotion, but also to anonymous citizens hunted in their private space by telemarketing, by the integrated circuit of cameras, by the control of communication flows, or by military observation systems. Media *voyeurism* and exhibitionism make a good couple: intimacy becomes a tradable commodity and modesty becomes a trite Victorian attitude.
>
> BENSAID, 2008, p. 68

Lukács once stated that in the nineteenth century, workers were exploited as producers and from the twentieth century on as producers and consumers. One of the most beautiful scenes that Italian cinema has ever produced is that of the film "The working class goes to paradise", by director Enio Petri, shedding light on the exploitation of workers as consumers as a result of the advancement of the techniques of commodification of life in the twentieth century. Near the film's end, the worker Lulu gets home and exudes a reflective, thoughtful air. In the scene, he says absolutely nothing, while the camera rotates in one of the house's rooms and shows a huge amount of totally superfluous goods and trinkets, many of them produced by US corporations, representatives of the "consumer society" that arrived in the hands of Italian workers in the 1960s. The scene leads the viewer to enter Lulu's brain and think something like this: "Why did I produce so much? Why did I increase the pace of my work? To buy these things with no social sense, totally superfluous? What did I work so hard for?" In this film we also witness a dialogue between Lulu and Militina, a worker who went crazy and was hospitalized. In one of these conversations, Militina questions the usefulness of the pieces they produced. For us, this is one of the faces of the alienation of labour: production with no social sense, without any rational and conscious human control, focused solely and exclusively on the manipulation of "consumers" with a view to the irrational expanded reproduction of capital.

As Argentina tried to be reborn from the ashes of the military dictatorship and the neoliberal avalanche during the great crisis of 2001, the people and the middle class shouted "Enough! Let them all leave!" In the cases of Bolivia and Venezuela, the people could no longer stand the expropriation of their natural resources and the increase in prices of essential goods. In Mexico, workers represent their struggles with the slogan "sin maíz no hay país" – "without maize, there's no country". It arose in the context of the creation of NAFTA and expresses the need for national sovereignty in relation to maize.

In the case of seed control, there is a clear class struggle: socialization of Creole seed as historical heritage *versus* the increasing proliferation of genetically modified seeds that generate dependence, destroy the environment, and strangle the autonomy of small producers. It appears in the anti-capital struggles waged by La Via Campesina[3] against large multinational corporations. It is the struggle between the seed as a commodity, private property, focused on the accumulation of capital and domination of small producers *versus* self-management, the production of healthy food, the preservation of the traditional knowledge of ancient peoples who survived by sharing their seeds, not to mention the proper use of natural resources.

In the 2000s, the rebellions against the privatization of public goods initially involved peasant communities in the cities of Cochabamba and Achacachi, in Bolivia in 2000, and in Arequipa, Peru, in 2002, to name but two among many. The indigenous uprisings triggered by the contestation of the gas exploration

---

3 The Via Campesina is an international peasant organisation composed by social movements and organisations from all over the world. The organisation aims to articulate the social mobilisation processes of rural peoples at an international level.

agreements resulted in the resignation of the President of Bolivia in 2003 and the election of Evo Morales in 2005.

This is the struggle of Latin America's social movements against an accelerated process of "destruction" of human beings and nature, public goods, and the nation. It can also be seen in the struggles of Chiapas, Guerrero, and Oaxaca (Mexico) against the destruction of corn by transgenic products, and in the threats of commodification of the region's aquifer reserves, as can also be seen in the struggles for free *babaçu* groves, against the installation of hydroelectric power plants, mining companies, etc., in the Amazon.

In Mexico, because of the spread of Bt transgenic maize, driven by multinational corporations, native and Creole varieties were contaminated, putting a stop to a 12,000 year-old project of natural improvement of corn seeds by Mexican indigenous peoples and peasants, the farmers of that country. This is the commodification of the biological resources of peripheral countries: in the case of Brazil, mainly the Amazon, biopiracy in indigenous lands, among other places, appropriation of resources that would be used exclusively by indigenous people, etc. (Shiva, 2001, Sevá, 2019).

In Brazil, the "destruction" of Eucalyptus nurseries owned by a company called Aracruz and the struggles between the Via Campesina and Syngenta Seeds (a Swiss multinational), among others, can be interpreted as coping tactics in this new phase of commodification of capitalism, mainly due to the destruction of Creole seeds, the role science and technology plays in the capitalist system and the invasion of multinationals. With Syngenta in Paraná, one reason was the development of illegal experiments in transgenic soy and corn, in the buffer zone of the Iguaçu National Park, a practice prohibited by the Biosafety Law.[4]

The genetically modified tree industry has also entered this wave of maximum profit in the shortest possible time. According to Carrere and Lovera (2006), from a profit-oriented industrial perspective, forests have been perceived as "disorderly" and "unproductive". For many years, forest scientists and foresters have been trying to "improve" them. The answer to this was to establish plantations of a single species, in straight and equidistant rows, to obtain the largest possible volume of wood per hectare. Thus, forests are being progressively replaced by monocultures producing wood, and this can be one reason behind attacks by social movements. It is evident that the possession and use of land by transnational corporations from the 1960s to now did not

---

4   It should be noted that the struggle between La Via Campesina and Syngenta resulted in the murder of the activist Keno and the subsequent creation of the Agroecology Center in Paraná with his name.

take place in a peaceful and idyllic way. The advancement of corporations was followed by murders, assassination attempts, and expropriation of small and medium farmers, landless squatters, indigenous peoples, and Quilombolas around the globe.

Resisting this process, the workers of Acre, gathering around the figure of Chico Mendes, adopted a technique in their strategy of struggle for land and life – the "draw", a technique that prevents the destruction of trees as they're being felled. In the Northeast, the Quebradeiras de Coco[5] fought and fight tirelessly for the Free Babaçu Law, to counter the constant enclosures by the region's farms. In the 1970s, the Chipko Women's Movement (India) emerged, an organization to which the renowned militant-researcher Vandana Shiva belongs. They adopted the tactic of tying themselves to trees to prevent their felling and to fight against the dumping of atomic waste in the region.

In the 1980s, in Brazil, the Movement of People Affected by Dams (MAB) emerged on the Uruguay River (SC), and later expanded to a nation-wide movement. It is a way for populations affected by large dams to respond to compulsory displacements, derisory compensation, degradation of quality of life, destruction of their communities and cultural identities, degradation of the environment and the material basis of their productive activities, among other factors. "Water for life, not for death", was the motto that emerged in MAB and soon spread around the globe.

As we saw in the previous section, during the corporate-military dictatorships in Latin America, especially at the end of the twentieth century, we saw the return of "primitive accumulation", the patenting of indigenous resources, the confiscation of land, land grabbing, and a law of survival of the fittest, all of which are compounded to to the previously described processes of commodification.

Vandana Shiva (2001) calls this new phase of commodification by the names of "new colonialism", "biopiracy", and "biocolonialism". For her, "while biodiversity and indigenous knowledge systems satisfy the needs of millions of people, new patent and intellectual property rights systems" threaten to appropriate "the vital resources and processes of knowledge of the Third World and convert them into a monopoly to the advantage of Northern companies. Patents are therefore at the centre of new colonialism" (Shiva, 2001, p. 320).

In large corporations, increased productivity of the workforce and planned obsolescence processes have risen to staggering levels. Innovation, as a form of

---

5 Quebradeiras de Coco (Coconut Breakers) are traditional gatherers specialized in harvesting goods from the Babaçu Coconut Palms.

capital accumulation and consumer domination, reaches a new level, which if not irrational, borders on irrationality. For instance, an employee of a large corporation in the children's food industry reports that "people are going crazy. They are trying to invent peanut butter with structures that pop like popcorn and make it so that it gives them energy, besides helping students focus at school" (Schor, 2009, p. 128).

The increase in productivity in agribusinesses, which gave rise to diseases such as mad cow, avian flu, and recently swine flu, are part of the process of intensification of goods that we are experiencing. Financial capital cannot wait and the meat commodity cannot wait and must be produced in the shortest possible time.

In the late 1970s – a period of "redemocratization", which Florestan Fernandes (1986) called the "institutionalization of the dictatorship" – new land struggles emerged, which led to the emergence of the MST (Landless Worker's Movement) in 1984. Fernando Henrique Cardoso (1995–2002) undertook a timid agrarian reform, especially after international pressure (due to the Corumbiara Massacre) and the PT governments were in charge of strengthening existing settlements rather than creating new ones.

During the course of the Lula government's eight years, agribusiness was strengthened, mainly with the policy of expanding sugarcane and alcohol production, and a great expansion of soybeans in the country. If Lula's administration had some contradictions, including strengthening some settlements, and in a more general way, trying to build a type of capitalism with some social rights, with the 2016 coup agribusiness got a "green light" to invade indigenous lands, destroy the forest, and move towards "virgin" areas. The Brazilian tragedy gained new airs with the (swindled) election of Jair Messias Bolsonaro (2019–2022), giving rise to new scenes of destruction of nature and human beings. While we were finishing this chapter, the Amazon burned in flames in the face of capital's insatiable greed. Brazil has become a protectorate of the USA, our bourgeoisie, associated and dependent, unconditionally supports and uses the retired captain. It needs to appropriate more land, in its incessant and uncontrollable march that destroys everything and everyone.

CHAPTER 2

Destructive Production and Agroecology

1 Introduction

This chapter aims to a) make a critique of the Green Revolution; b) analyse the role of agroecology as a "way out" of the agricultural crisis, from the point of view of social movements; c) note the need to increase the number of state technicians (such as those employed by the Brazilian states' agricultural technical assistance companies, known as Ematers, among others) and a new methodological approach and a political project for rural extension qualitatively distinct from the standard that still exists; d) the indispensable role of the social movements of the countryside, to wage the class struggle and take over land and achieve the public policies necessary for the "development of the countryside".

We focus on the work of Enio Guterres (2006) mainly because of the criticism handed out by him to the current rural extension model in the country. As reported in the Introduction of Guterres (2006), he graduated in agronomy, worked at Emater (RS), was a member of the PT, the MST and the Via Campesina. He died very young, but still managed to bring together "scientific" concerns, studying and researching together with social militancy, always standing alongside and together with rural workers and peasants. A militant scientist within a revolutionary tradition. He was both a pedagogue and a technologist at the same time.

We will also analyse the contributions of the Chilean researcher-extensionist Miguel Altieri, PhD in entomology and professor at the University of California, Berkeley. He has long lived in the USA, but constantly visits Brazil, Chile, and other Latin American countries, influencing agroecological debates and practices. Besides these, we tried to summarize the main ideas of sociologist Sevilla Guzmán and her research group at the University of Córdoba (Spain), which also greatly influenced Brazilian agroecology researchers-extensionists, including being a doctoral advisor to many of them, as well as acting directly in relationship with some social movements.

As agroecology is a very new area, we also chose to read some articles by Francisco Roberto Caporal and José Antônio Costabeber. Caporal is an agronomist, Master in Rural Extension (CPGER/UFSM), PhD at the Agroecology, Peasantry and History Program (University of Córdoba – Spain) and participant in the rural extension of Emater/RS-Ascar. He participated in the Ministry of Agrarian Development (Lula Government – 2003–2010). Costabeber is an agronomist, master in Rural Extension (CPGER/UFSM), who also has a PhD in Córdoba and works as a rural extensionist at Emater/RS.[1] He is currently at the Federal University of Santa Maria (UFSM).

Pinheiro Machado graduated in Agronomy (1950), and got his PhD in Plant Sciences (1959) from the Federal University of Rio Grande do Sul (UFRGS). He is a retired professor of the Graduate Program in Agrosystems at the Federal University of Santa Catarina (UFSC). Despite writing very little, Pinheiro Machado is a Latin-American reference in the field of agroecology.[2] For the critique of intellectual property, we chose to rely on the texts by Vandana. According to information from the internet, Vandana Shiva is a physicist and ecofeminist, as well as being the director of the Research Foundation for Science, Technology, and Ecology in New Delhi. For her, "a very long name for a very humble goal, which is to put research effectively at the service of popular and rural movements, and not just pretend that we are helping them".

As we will see in this chapter, in the 1970s she took part in the Chipko Women's Movement, formed mostly by women who adopted the tactic of tying themselves to trees to prevent their felling and the dumping of atomic waste in the region. She is the leader of the International Forum on Globalization and won the Right Livelihood Award in 1993, considered an alternative version of the Nobel Peace Prize. Besides these, we also rely on the writings and action of a research-extension group that entered the debate more recently: Grupo Terra e Liberdade (UFSM), which has been theorizing about agroecology and advising social movements.

Here, we chose to take into account a larger range of researchers-extensionists to contrast different trends in the field of agroecology, the points of contact between their arguments and the contradictions that appear. To begin with, let's look at some criticism of the "Green Revolution".

---

1 Rio Grande do Sul's Public Technical Assistance and Rural Extension Company aims to carry out agricultural research, provide technical assistance and create and adapt technologies using educational and participatory methodologies.
2 According to an e-mail Pinheiro Machado sent us: "Dear Henrique; Unfortunately I study a lot and write too little. … That is why my texts are rare, although I have a lot of experience in the area you are studying. I have a book where you will find a lot of material for your thesis: Pastoreio Racional Voisin, Editora Cinco Continentes" (Pinheiro Machado, 2009). After this letter was sent, Pinheiro Machado's book also came out by publisher Editora Expressão Popular.

2 Green Revolution or Green Con? The Advancement of Destructive Forces in the Countryside

In agribusiness, the obscuring of domination in the countryside occurs in the context of the so-called "Green Revolution" – or is it a Green Con?.[3] According to Guterres (2006), there was an euphoria around transgenics in the 1970s/1980s, when small farmers, bewitched by the many wonders of the "green gold" replaced even their orchards and vegetable gardens to "make money".

Recovering a bias dating to the Enlightenment, the ideology and practice of technical progress in agriculture become the newest form of domination exercised by capitalism. What we could call "the political economy of the Green Revolution" or "the political economy of the Green Con" can be described as follows: concentration of land, mergers and acquisitions in the seed business, almost complete domination of production and distribution by some large corporations in northern countries, lack of autonomy for producers, predominance of financial capital, etc. There was also a drastic restructuring of the input production and industrial transformation sectors, financing and credit institutions and mechanisms, marketing circuits, and market structure (Costa Neto, 1999).

Educational, research, and technical assistance institutions were also readjusted to this model, with a view to training researchers, specialists, extension workers, and other professionals within the philosophy of the Green Revolution. From a historical perspective, Pinheiro Machado observes that:

> since Liebig in 1848, the capitalist industry has seen agriculture as an excellent source for the reproduction of capital, and, from there, agronomy schools around the world do nothing other than teach students to apply synthetic fertilizers, pesticides, and heavy machinery.
> PINHEIRO MACHADO, 2009

Evangelista also highlights the massive investment in advertising to "convince" or manipulate consumers. According to this journalist, reflecting on the most recent period:

> In an unprecedented decision, Mendocino County, California, USA, managed to ban Genetically Modified Organisms (GMOs) by popular vote. It

---

3 T.N. – The Portuguese "golpe" has a polysemantic meaning here, as both a con and a coup.

is estimated that the industry spent over $600,000 on radio, newspapers, and TV ads in the campaign against the ban.

Social movements expect Mendocino's decision to be repeated in other locations, given the precedent set.

Previously, the biotechnology industry had already promoted another advertising blitz, this time in the State of Oregon, also in the USA.

The State intended to impose the labelling of transgenic foods and, before the campaign, had the support of 70% of the electorate. After the industry spent $5.53 million on advertising, stating that labelling would make food more expensive, most of the population voted against it.

The campaign, entitled "Coalition against the costly Labelling Act", was sponsored by DuPont, Syngenta, Dow Agro and Monsanto.

A similar advertising campaign also took place in Brazil. In advertisements published in the press, Monsanto spent R$6 million trying to sway public opinion with the phrase: "If you have thought of a better world, you have thought of transgenics".

EVANGELISTA, 2004, s/n.

Beyond the use of massive propaganda, in states where properties remained small, producers, supposedly "autonomous" and "independent", became true "slaves" through outsourcing processes driven by large corporations, but with the difference of being "owners of the means of production". According to Dowbor:

> It is very important to monitor in agriculture a form (...) of externalization of production in relations with large agro-industrial companies, such as Batavo, Parmalat, Sadia, Souza Cruz, Cica and others. Basically, it is a question of promoting the production of small producers who will work according to extremely strict technical specifications of the company that runs the technical chain in a region and often supplies the raw material itself. Although they represent a monopsony in terms of a single buyer, with no alternatives for small producers, these companies sing loas to market mechanisms, forcing small producers to compete. The practical result is a form of proletarianization of producers who own their means of production. The sparse data that reaches us show that a milk producer receives less than 10 cents per litre produced, that tobacco producers receive less than half a cigarette's worth from each pack sold, and so on. Here, any drop in the market does not generate accumulation of inventories in the final producer, but a reduction in orders from small producers, who will bear the impact of the crisis. Thus, risk capital

is generated coupled with a powerful mechanism for transferring risk to the workers themselves.

DOWBOR, 2001, s/n

The consequences of the "Green Con" are too well known to be detailed. It is up to us only to highlight them: increased unemployment; land tenure concentration; soil degradation; compromised quality and quantity of water resources; devastation of forests and native fields; impoverishment of genetic diversity of cultivars,[4] plants and animals; contamination of food consumed by the population; increased allergies, deaths or disability; increased commodification of rural land[5] etc.

The tendency towards monoculture, caused by crop specialization and the importation of predominantly chemical fertilizers from third parties, causes the unambiguous exploitation of humus resources, that is, the stable fertility of the land, and the abnormal development of parasitic insects. It also intensifies dependence on the market, increases the costs of transporting products, leading to deterioration and the need for chemical conservation (Costa Neto, 1999).

The logic of maximum profit in the shortest possible time, always present throughout the history of capitalism, has reached new heights over the last 50 years. Increasingly guided by the "perfect tripod" or by married purchases of seeds, herbicides and machines, the agro-industrial structure has been overhauled, consolidating a "structure of power and domination".

The genetically modified tree industry has also entered this wave of maximum profit in the shortest possible time (Lang, 2006). According to Carrere and Lovera (2006), from a profit-oriented industrial perspective, forests have been perceived as "disorderly" and "unproductive". For many years, forest scientists and foresters have been trying to "improve" them. The answer to this was to establish plantations of a single species, in straight and equidistant rows, to get the largest possible volume of wood per hectare. Forests are being progressively replaced by wood production monocultures.

Carrere and Lovera (2006) remind us that different measures have been taken to "improve" forests. The first step was to research which trees were appropriate for each environment. We must not forget that FAO (The UN

---

[4] See, for example, the terminator gene, which causes the seed harvested by the farmer to "kill itself" when trying to sow it again.
[5] Sevin is a case in point. A pesticide produced in Bhopal (India), it led to a disaster caused by the leakage of a gas used in its production. The episode caused the death of thousands of people and incapacitated more than 400,000. Furthermore, the advancement of hybrid seeds has already led numerous Indian producers to suicide.

Food and Agriculture Organization) played an important role in this process, particularly in the case of eucalyptus. It also supported the use of the entire "Green Revolution" package: mechanization, use of herbicides, etc.

The specificity of the Latin-American case is that the Green Revolution was implemented in a process of counter-revolution started by the military dictatorships to recompose the power structure in "rural" Latin America – in the Brazilian case, with its roots archaic Brazil, from backward Brazil, the Brazil of landowners with properties the size of countries. Large landowners, representatives of the Latin America of the past, banded together with the military and portions of the industrial bourgeoisie, the church and the CIA, to carry out the coups that stopped the advance of socialism in the region.[6]

The discourse of large corporations, as always, was that this revolution would "solve the people's food problems". For us, food, like any commodity, is use-value "impregnated" with exchange-value in an exploitation-based society. Behind soy, for instance, there is a large industry that makes a lot of money. Creating general production conditions in its favour, it is estimated that, in 2003, about US$84 billion reached the seed, pesticides, fertilizers, machines, implements, fuels, transportation, storage, security, financial intermediation, processing and packaging businesses, among others. (Guterres, 2006).

Many small farmers, spellbound by the promise of easy profit, perhaps due to huge investments in advertising, ended up "getting tied up in monoculture as if they were large producers" (Guterres, 2006). This generated, a rural exodus and huge social and environmental costs, as pesticides affect producers and people, contaminates water and soil and reduces forests; and, consequently, reduces water and affects the climate, changing it with 'veranicos' (dry spells in the rainy season), drought and out-of-season cold and heat waves (Guterres, 2006). Rural extension or "technology transfer" – topics to which we will return in the following sections – was the main vehicle to boost the industrialization and technification of agriculture in the United States and Europe and the so-called "Green Revolution" in countries of the Global South.

The technician should be an expert with mastery of techniques and practices and good persusasion skills (technical assistance). Rural extension was presented as an informal education process to improve the economic and social conditions of rural producers. The "extension agent" defined socioeconomic factors as major components for action, working for the development

---

6 On this, besides the debate on the Peasant Leagues, the growing unionization in the field, the conservative unions created by the church, the role of the PCB etc., see Dreifuss (2008), Santos (2000), Umbelino (2005), Rodrigues (2005), Oliveira (2009), and Novaes (2008).

of agriculture and technological innovation in agricultural production (Guterres, 2006).

Monocultures create more and more pests and dramatically increase problems with insects, fungi and so-called weeds. GMOs decrease these problems for a few years, but then they come back stronger, increasing farmers' dependence on corporations. Embrapa's transgenic beans contain a Brazil nut gene that, when tested in the USA, caused allergic reactions. A laboratory in York, in the United Kingdom, found that soy allergies increased by 50% in that country after the commercialization of transgenic soy (Guterres, 2006).

Defending agroecological principles, Altieri (2012) criticizes the "Green Revolution", showing that its "benefits" were extremely unequal and that it also contributed to the dissemination of environmental problems, such as soil erosion, desertification, pollution by pesticides and loss of biodiversity. In a sense, this reveals a failure of the dominant development "paradigm". Conventional development strategies proved to be fundamentally limited in their ability to promote equitable and sustainable development, says this researcher.

The conventional approach did not address the ecological causes of environmental problems in modern agriculture, deeply rooted in the monoculture structure prevalent in large-scale production systems. The advocates of the Green Revolution cannot and do not want to recognize the fact that their model's limiting factors are only the symptoms of a more systemic disorder, inherent to imbalances within this agroecosystem. On the other hand, approaches that perceive the problem of sustainability only as a technological production challenge cannot reach the fundamental reasons behind the unsustainability of agricultural systems, adds Altieri.

After systematizing the criticism of these agroecology researchers-extensionists to the Green Revolution, let's try to dwell on a theme derived from the Green Revolution: the role of patents in new colonialism.

3   Patents as a New Form of Colonialism

To develop this section, we rely on the arguments of Vandana Shiva (2001). She calls this new phase of commodification of capitalism "new colonialism", "bio-piracy" or "bio-colonialism". According to this militant researcher:

> "while biodiversity and indigenous knowledge systems satisfy the needs of millions of people, new patent and intellectual property rights systems" threaten to appropriate "the vital resources and processes of knowledge of the Third World and convert them into a monopoly to the

> advantage of Northern companies. Patents are therefore at the centre of new colonialism".
>
> SHIVA, 2001, p. 320[7]

If a recurring word in patents in the fifteenth century was "discover and conquer", in the twentieth and twenty-first centuries the name recolonization is predominant:

> religion is no longer a primordial justification for the current conquest [as it was in the past]. Colonization is a "secular" project, but there is a new market religion that drives this secular project. Territory, gold and mineral resources are no longer the targets of conquest. Markets and economic systems are the resources to be controlled. Knowledge itself has to be converted into property, analogous to what took place with lands during colonization.
>
> SHIVA, 2001, p. 321–322

For her, the decisive milestone in the advancement of patents is the Trips (Intellectual Property Rights) agreement in the Uruguay Round in1992, prepared by an industry coalition and the Intellectual Property Committee (IPC). Prior to the Uruguay Round, intellectual property rights were not covered by GATT (which became the WTO). The set of patentable materials has been expanded, removing all limits to what is patentable.

Bio-piracy is the process of patenting biodiversity, constituent parts of living beings and products derived from them, based on indigenous knowledge while excluding indigenous people. Patents are a right to a monopoly, which excludes other companies from the production, use, sale or import of products that are patented or products manufactured through a patented process. Shiva notes that often patents claiming to be "innovations" already exist in the knowledge systems of indigenous communities.

For this reason, patents based on bio-piracy not only deny the collective innovations accumulated over time and the creativity of peoples, but also become a system of "enclosure of intellectual and biological common goods that are essential to their survival" (Shiva, 2001). Shiva also notes that

---

7   For a retrospective of patents since the 15th century, see Shiva (2005) and Barbosa de Oliveira (2005). For further details, see also Andrioli and Fuchs (2007).

traditional knowledge moves to large corporations via piracy and returns to workers in expropriated countries as expensive, commodified products, etc.[8]

We believe that the fight against bio-piracy and against Trips emerged as one of the central elements of the current anti-globalization struggle. It includes movements led by indigenous communities, farmers, women, as well as ecological and community health movements. This is one of the few areas in which "third world countries have resisted the Northern hegemony, which has made the revision of Trips one of the most significant stages of North/South conflicts" (Shiva, 2001, p. 325).

The challenges for a "post-globalisation agenda" can be expressed in the two currents of the Movement for Living Democracy (India). A more radical trend challenges the commodification of life, inherent in Trips, the WTO and the erosion of the cultural and biological diversity of bio-piracy. For this current, "resisting bio-piracy is resisting colonization". The other trend is more technocratic – always in Shiva's words – and aims at a correction within the commercial and legal logic of the commodification of life and of the monopolies on knowledge. In this case, the keywords are bioprospection and sharing of benefits, that is, the idea that those who claim patents on indigenous knowledge should share the benefits of the profits of their monopolies with the original innovators (Shiva, 2001, p. 329). The author makes a variety of criticisms of this current.

Shiva highlights some challenges for a "post-globalisation agenda", including the protection of creators' rights and self-management. She takes up a phrase from the liberation movement in India: "self-government is our birth right". For her, "self-government does not imply governance by a centralized state, but by decentralized communities". "In our village, we are the government" is a slogan of the grassroots environmental movement, referring to the right to local sovereignty (Shiva, 2005).[9]

She believes the movements against bio-piracy and Trips also formed a "new pluralist politics, a rainbow politics, with the generosity and the ability to include a space for the struggles of indigenous communities and to defend local sovereignty, as well as for movements that fight for the satisfaction of

---

8 The international network of warriors against patents emerged as a result of the Neem campaign in India (see Shiva, 2001). Raw (2000) and Garcia dos Santos (2005) report the famous Novartis-Bioamazônia (Brasil) case.
9 At times, Vandana Shiva's theory dangerously approaches the very fashionable trend of local development, without making the proper connections with national debates. About this, see Sampaio Jr. (2006) and Montaño (2004).

basic needs and the defence of national sovereignty" (Shiva, 2005). She concludes by saying:

> Resistance to bio-piracy is resistance to the definitive colonization of life itself – the future of evolution, as well as the future of non-Western traditions of knowledge and relationship with nature. It is a fight to protect the freedom of evolution for different species. It is a fight for the conservation of cultural, animal and plant diversity.
> SHIVA, 2005, p. 328

After portraying the role of patents in this new phase of capitalism, let us now see what experiences and what historical periods agroecology researchers-extensionists rescue in their efforts.

4 Rescue of Historical Experiences of Alternative Agriculture: Clues to the Understanding of Agroecology

Among the researchers under study, Professor Pinheiro Machado seems to be the one who is most committed to the dissemination of theorists and historical experiences of "alternative" agriculture.

In this section, we will mainly rely on some books that he suggested should be translated and published in Brazil. One of these books is written by Francis Chaboussou, a French researcher. For Pinheiro Machado, the book helps us understand the true and complex process of protecting plants from the harmful action of parasitic agents: insects, fungi, bacteria, viruses, mites, nematodes and coccids (Pinheiro Machado, 2009).

> Everyone should read and meditate on this text: producers, to question their technicians when they recommend pesticides and/or soluble fertilizers; students, to ask their teachers about Chaboussou's positions; technicians, to train themselves in poison-free production practices; teachers, to lead their students to a position contrary to conventional agronomy; and, finally, researchers who have distanced themselves from reality, who should step down from their fragile pedestal and come to the level where life takes place and, therefore, where the truth lies.
> PINHEIRO MACHADO, 2006, p. 16

In the 1970s, Chaboussou wrote about the theory of trophobiosis, a pillar of agroecology. It forms the basis on which the production of clean and healthy food can be built, dispensing with the use of pesticides and chemically synthesized soluble fertilizers (Pinheiro Machado, 2006). Soluble fertilizers and pesticides attract parasites, thus generating a cycle of dependence. The primary objective of farmers should then be to protect plants from the action of parasites (Pinheiro Machado, 2006).

Before Chaboussou, those concerned with clean animal and plant agricultural production did not have access to an ancient practice – and its theoretical support – known and disseminated by true agroecologists: plants that grow in soils that are rich in organic matter originating from manure are not attacked by pests and diseases (Pinheiro Machado, 2006). Chaboussou brought attention to the fact that pesticide use leads to the emergence of new diseases. The study of biological imbalances produced by different conventional practices demonstrates that, rather than controlling parasites, they cause a disturbance in the physiology of plants, thus aggravating the problem, transforming beings that previously kept a harmonious coexistence with plants into parasites (Pinheiro Machado, 2006, p. 12–13).

They are "iatrogenic" diseases, that is, diseases caused by the use of supposed medicines. It is not by accident that the few dozen pests and plant diseases known just over half a century ago today reach the thousands (Pinheiro Machado, 2006). Pinheiro Machado observes that the warnings of Howard, Russel, Rusch, Voisin, Faulkner and many others have been underestimated by mainstream science (Pinheiro Machado, 2006). He believes that there is a powerful game of interests in the agro-industrial sector, and the price is being paid in various ways by agricultural producers, small, medium and large (Pinheiro Machado, 2006, p. 13).

Chaboussou identified the root causes of the problem, and he proposes a solution applying the correction of mineral deficiencies in the soil, especially microelements (Pinheiro Machado, 2006). The primary cause of parasitic infections is nutritional imbalance. The equilibrium of the mineral composition of the soil is the *sine qua non* of its fertility. The problem is *how* to achieve this balance (Pinheiro Machado, 2006).

Pinheiro Machado's main objective is to adopt techniques to detoxify soils attacked by predatory agriculture. He states that we must adopt a holistic view, always working with the causes, and not on the effects. He believes that the theory of biocenosis, based on substantial experimental results, alongside Chaboussou's theory of trophobiosis, make up the foundation of a new and instigating paradigm, free from pernicious economic dependencies, recovering the dialectical, or we could even say true, sense of the timeworn expression "working with nature". In a prophetic tone, he concludes:

> For scientists without prejudice and farmers-researchers, this is an open door to building a doctrine that offers producers the technology of life, in which the wonderful harmony of nature with "its own consciousness, that is, humans" would be achieved. This construction will be completed when science can develop a production model capable of feeding humanity without the dilapidation of non-renewable resources, through the wonderful work of soil life, in harmony with the maximum capture of solar energy through photosynthesis.
> PINHEIRO MACHADO, 2006

Clean, profitable and sustainable agriculture is built from an understanding of the indispensability of using solar energy inputs and soil life dynamics, that is, this is how true agroecology is put into practice – a safe way to perpetually produce clean food, as demanded by the very survival of humanity (Pinheiro Machado, 2006, p. 17). Altieri has carried out many historical studies aiming to recognize the "millennia-old knowledge" and "repositories of genetic diversity" of Peruvian and Mexican societies. He believes the knowledge of Peruvian and Mexican farmers cannot be neglected.

In traditional agroecosystems, the predominance of complex and diversified cultivation systems is of paramount importance for peasants, as the interactions between cultivated plants, animals and trees result in beneficial synergies that allow agroecosystems to promote their own soil fertility, pest control and productivity (Altieri, 20012).

Peasants working with traditional production systems have sophisticated knowledge and understanding of the agricultural biodiversity they handle. This is the reason why agroecologists oppose approaches that separate the study of agricultural biodiversity and the study of regional cultures.

It is noteworthy that researchers only recently began to describe and record part of this knowledge, hitherto ignored by "conventional" science (Altieri, 2012). When faced with specific challenges, such as slopes, floods, droughts, pests, diseases and low soil fertility, small farmers all over the world have developed particular work systems to overcome them (Altieri, 2012).

Costa Neto (1999) notes that researchers from the 1920s already sought references in Hindu, Inca, and other alternative agricultures and opposed "conventional agriculture". He believes agroecological systems are an "alternative to monoculture, to business agriculture, which sees agricultural exploitation as a business, an endeavour that must have its profits maximized and its losses minimized", made through a large pact made "between national and international big capital".

Among the precursors of agroecology, Costa Neto highlights Rudolph Steiner (1861–1925), Albert Howard (1873–1947) and Claude Aubert and notes that these intellectuals help us to "demystify now-consolidated concepts". He notes that researchers in this field are not nostalgic, but intend to rework some principles that industrial agriculture has cast aside (Costa Neto, 1999). After taking a brief look at the theorists and historical experiences "recovered" by agroecology researchers-extensionists, let's try to deepen the debate on the necessary topic of the agroecological transition.

5 The Concept of Agroecology and the Need for an Agroecological Transition

In this section, we intend to address how agroecology was conceived and the arguments used for the transition from one paradigm to another. We synthesize the ideas of Caporal and Costabeber, Guterres, Altieri, Sevilla Guzman and Costa Neto. According to Guterres (2006), one of the biggest expropriations that multinational corporate agriculture inflicted upon peasants targeted centuries of knowledge that were transmitted from generation to generation, especially through oral traditions and experiences – teaching and learning through practice. Much of this knowledge was not formally recorded, it was not written down. Some have been lost forever. It is necessary to regain this lost heritage and seek new knowledge made possible thanks to constant new advances in human knowledge, based on the agroecological principles of production.

He observes that seeds are basic inputs that should be under the control of farmers and their organizations. Harvesting, selecting, conserving, experimenting, crossbreeding, and improving seeds and seedlings should be practiced by farmers to build a new model in agriculture.

Agroecology has become part of the vocabulary of social movements both because of the desire to produce healthy foods, and also because of the enormous costs that conventional agriculture has entailed.[10] The vast majority of fertilizers' prices are dependent on the price of oil, which peaked between 1998 and 2008 and led many to adopt the agroecological matrix by need, not by an abstract desire.

Sustainable agriculture generally refers to a way of practicing agriculture that seeks to ensure sustainable productivity in the long term through the use

---

10   Costa Neto (1999) briefly recovers the "evolution" of the technological models used in the MST settlements. See also Christoffoli (2009).

of ecologically safe management practices. This requires seeing agriculture as an ecosystem (hence the term agroecosystem) and that agricultural practices and research are not concerned with high levels of productivity of a particular product, but with the optimization of the system as a whole. This also requires considering not only economic production, but the vital problem of ecological stability and sustainability (Guterres, 2006).

Guterres (2006) also highlights that a desire to leave chemical agriculture behind and produce without poisons and chemical fertilizers, adopting an ecologically based technological model, grows day by day among small farmers. In other words, there is an attempt to reduce farmers' dependence and increase their autonomy to build a new way of producing.

Guterres is against a radical transition. For him, changing everything at once can easily go wrong. This is because we do not have enough technical assistance and research in the agroecology in Brazil to support all small farmers who might want to start transitioning to agroecology. The first steps to be taken are caring for and recovering natural fertility and ecological management of the soil. Among the advantages of this practice, he highlights less costs with fertilizers, greater ease to control competing plants, lower transfer of income to fertilizer factories, greater autonomy for the farmer, greater resistance of plants in periods of drought and a greater use of waste – such as manure, debris, bagasse etc. – on the property.

The gradual and partial replacement of inputs importation for production will require, for instance, the internal production of inputs, including native and Creole seeds, organic and green fertilizers, and pest and disease management practices. It is possible to decrease and gradually eliminate the use of poisons in agriculture as entire communities move together to another technological model, based on the diversification of production (Guterres, 2006).

According to Gloria Guzman Casado, this model should have the objective of producing nutritional, high-quality food in sufficient quantities and working with natural systems rather than claiming to dominate them. In an associated, cooperative way, the production, transportation, storage, industrialization and commercialization of production infrastructure must be built to create what we call "new general production conditions" for this new system. This will make farmers independent of middlemen, who currently take the majority of their income (Guterres, 2006).

According to Altieri, through the use of appropriate technologies, experimentation and implementation of organic agriculture and other techniques with low input use, one can ensure that alternative systems result in a strengthening not only of some families, but of entire communities. Thus, technological interventions and processes must be complemented with education programs that preserve and reinforce a peasant rationale while helping to transition to new technologies and relations with the market and social organizations – for instance, through agroecological programs promoted by NGOs (Altieri, 2012).

He also warns us that many NGO projects based on an agroecological approach lack formal and detailed assessments.[11] However, there is strong evidence that many of these organizations have generated and adapted technological innovations capable of significantly contributing to improving the living conditions of peasants, increasing their food security, strengthening subsistence production, generating sources of income and improving the natural resource base.

These programs were successful through new technologies and institutional arrangements, as well as the use of original methods of promoting the participation of rural communities (Altieri, 2012). He believes that we have few adequate instruments or indicators to assess the feasibility, adaptability and durability of agroecological programs. He does recognize two relatively new procedures as "promising": rapid participatory diagnosis (DRP, from the acronym in Portuguese) and natural resource accounting (CRN, same).

Rapid participatory diagnosis techniques emphasize non-formal methods of data collection and presentation, aiming to favour a participatory process between local people and researchers. To conduct a DRP, a multidisciplinary team works with the local community in a series of steps, starting with the choice of location and ending with the evaluation and monitoring of the project.

For Altieri (2012), the objective is to mobilize communities to define priority problems and opportunities, preparing specific intervention plans in chosen locations. Data collection and presentation is a complex process that uses maps, diagrams, timelines, and individual and group semi-structured interviews. Potential technologies are assessed through very general criteria, based on environmental, economic and social concerns expressed by local residents.

To summarize his ideas, we can say he believes that the development and diffusion of agroecological technologies and the promotion of sustainable agriculture require changes in research agendas, as well as agrarian policies and economic systems covering open markets, prices, as well as government incentives (Altieri, 2012).

Bringing the debate to contemporary issues, Altieri notes that the agroecological approach is also more sensitive to the complexities of local agricultural systems. In it, performance criteria include not only increasing production, but

---

11   For a critique of NGOs and, more generally, the third sector, see Montaño (2004).

also variables such as sustainability, food security, biological stability, resource conservation and equity. He concludes:

> A problem with the Green Revolution in heterogeneous agricultural regions is that it focused its efforts on the farmers with the most resources, those at the top of the gradient, hoping that "progressive or advanced farmers" would see themselves as an example to others in a diffusionist process of technology transfer. Agroecologists, on the contrary, emphasize that for development to be really bottom-up, it must start with those small farmers at the bottom of the gradient. Thus, the agroecological approach has proven to be culturally compatible, as it is built on traditional agricultural knowledge, combining it with elements of modern agricultural science.
>
> ALTIERI, 2012

The resulting techniques are also ecologically correct, as they do not radically change or transform the peasant ecosystem, but rather identify traditional and/or new management elements that, once incorporated, optimize production. The emphasis on local resources reduces production costs, making agroecological technologies economically viable (Altieri, 2012).

By definition, agroecological production models and techniques lead to higher levels of participation. In practical terms, the application of agroecological principles to rural development programs has been translated into a diversity of research and demonstration programs and alternative production systems. These programs have several objectives: a) to improve the production of basic food at the level of production units, strengthening and enriching the diet of families. This has involved a renewed appreciation of traditional products and conservation of the germplasms of local varieties; b) the rescue and reassessment of peasant knowledge and technologies; c) promoting the efficient use of local resources (i.e., land, labour, agricultural by-products, etc.); d) an increase in plant and animal diversity in order to reduce risks; e) improvements of the natural resource base through the conservation and regeneration of water and soil, emphasizing the control of erosion, water collection, reforestation, etc.; f) reducing the use of external inputs, thus reducing dependence while sustaining productivity levels (Altieri, 2012).

According to Guterres (2006), agroecology is a way of understanding and acting to "enpeasant" agriculture, animal agriculture, forestry and agroextractivism, based on intergenerational (non-exploitation of children and the elderly), class (non-exploitation of labour by capital), species (non-exploitation of natural resources), gender (non-exploitation of women by men) and identity (non-exploitation between ethnicities) consciousness.

Caporal and Costabeber state (2002):

> agroecology brings us the idea and expectation of a new agriculture, able to do good to people and the environment as a whole, moving away from the dominant orientation of an agriculture intensive in capital, energy and non-renewable resources, which is environmentally and socially aggressive and causes economic dependence.

Caporal and Costabeber defend agroecology as a scientific paradigm, within a multidimensional analysis that aims at the transition to sustainable rural development. They believe agroecology encompasses "less aggressive agricultural styles that promote 'social inclusion'", creating a "new agriculture capable of doing good to people and nature". It brings together several fields of knowledge to make up its theoretical and methodological field. Agroecology is the science that lays the foundations for building different styles of sustainable agriculture and "sustainable rural development" strategies. It is based on the concept of the agroecosystem as a unit of analysis, ultimately aiming to provide the scientific foundation (principles, concepts and methodologies) to support a process of transition from the current model of conventional agriculture to different styles of sustainable agriculture (Altieri *apud* Caporal and Costabeber, p. 71–72). The basic principles of a sustainable agro-ecosystem are: the conservation of renewable resources, the adaptation of crops to the environment and the maintenance of a moderate but sustainable level of productivity and the diversification of crops, among others. (Altieri, 2012, p. 65).

However, Caporal and Costabeber warn that the simple replacement of agrochemicals by poorly managed organic fertilizers may not be the solution and may even cause other types of contamination. Costa Neto points out the importance of understanding, even if only incipiently, to what extent agroecology performs the role of a "counter-science", or "alternative science", in the contemporary scenario. Interdisciplinary par excellence, it could, over time, consolidate itself as a field for analysis, research and verifiability around which alternative knowledge in agriculture could be organized.

Although permeated by contradictions, Costa Neto (1999) points out some characteristics to try to distinguish alternative and industrial agricultural paradigms. The first is predominantly characterized by polyculture, local and regional markets with technological autonomy and little waste, natural processes used for long-term fertility, and social, economic and ecological stability. Industrial agriculture is based on monoculture, using varieties selected for high yield, large industrial companies and wage labour, distant markets, technologies with large waste and non-renewable energies (oil, atomic) as well as fossil fuels and chemicals, being therefore ecologically, economically and socially unstable.

According to Costa Neto, intensive, conventional agriculture based on the "Green Revolution" model can be described by a set of characteristics. Regarding its focus, it is reductionist, not systemic; from the point of view of its objectives, it acts in the short term, with a productivist conception, emphasizing physical performance and not taking "environmental costs" into account. It is highly unstable and is based on simplified systems with low diversity. As far as its techniques, it uses synthetic fertilizers, practices intensive land use, permanent agriculture, and chemical pest control and adopts transgenic plants for pest control (Costa Neto, 1999).

Agroecological agriculture has a holistic approach, the use of a systemic perspective and an emphasis on interrelationships. The goals are long term. It is based on the agroecosystem, incorporating "environmental costs". It is based on complex systems with high diversity and is thus considered stable. The techniques it adopts include organic fertilization, nutrient recycling, soil conservation, crop and animal rotation, polycultures, integrated and biological pest management and biodiversity management to control these. According to Costa Neto, agroecology takes lessons from several scientific disciplines and intends to study agrarian activity from an ecological perspective. For him, "conventional" science has a reductionist approach to science, fragmented and Cartesian.

Organic agriculture is agriculture that largely avoids or excludes the use of synthetic fertilizers and pesticides. Whenever possible, external resources, such as "chemicals" and commercially acquired fuels, are replaced by resources found at or near the agricultural production unit. These internal resources include solar or wind energy, biological pest controls, biologically fixed nitrogen, and other nutrients released from organic matter or soil reserves. The specific options on which organic farming is based, as far as possible, include crop rotations, use of crop residues, animal manure, legumes and green manures, as well as of residues external to the production unit, mechanical cultivation and ground rocks containing minerals, etc. (Costa Neto, 1999).

All these practices lead to an increase in soil organic matter, the elimination of potentially toxic pesticide residues, the biological suppression of pests, adventitious diseases and weeds and the storage of rainwater, avoiding unnecessary runoff.

6 The Technical Assistance Required for Agroecology

This section aims to discuss the technical assistance and rural extension envisioned by agroecology researchers-extensionists, especially Enio Guterres. He proposes that a profound change in technical assistance and rural extension focused on agroecology must start by criticizing the current model, addressing the rural extension adopted – from "outside to inside" and from "others to someone", the basis of the diffusionist model. Then, one must point out the need for a radically new approach to extension, which seeks not to transfer technologies, or even "learn from farmers", but "to strengthen the already-extant capacity to generate knowledge, in the community – the ability to question, analyse and test possible solutions to their own problems" (Roger, 1987). Roger calls this "third generation" extension, contrasting it with the models of the "first generation" (directive) and the "second generation" (reactive, "farmers first") (Guterres, 2006).

Second-generation extension workers ask farmers to identify their problems and then leave to seek solutions, usually coming back soon after with the answers. After the extension worker brings the answer, there is little opportunity of choice for the farmer. As a result, Guterres (2006) starts from two premises: a) knowledge cannot be transferred; a person cannot learn the knowledge of another: he can only create his own. Learning is an active process, carried out by the one who learns, and not by a passive reception of knowledge "transmitted" to him.

In all these years, technical assistance and rural extension have always been detached from research, even in institutions responsible for both activities. There are many technologies in research establishments that do not reach farmers or are not suitable for small farmers because they are done in isolation, separate from reality. Guterres (2006) notes that Brazilian rural extension must generate concrete responses to the challenges of small farmers in Brazil – not as a vehicle for transmitting technological research results to farmers, but to strengthen the self-learning capacity of peasant families for self-management of agro-ecosystems and rural communities for sustainable rural development. Currently, Ater (Technical Assistance and Rural Extension) is insignificant in view of the demand necessary for the universalization of this service. Some even say that the number of technicians should double.

Conventional agronomic approaches to agricultural activity are based on the segmentation and division of scientific knowledge. Agronomy, as a scientific discipline, has the same defects as conventional science. It is: a) axiomatic – in which the very motivations of agrarian activity are not subject to discussion (profits and domination of producers); and b) productivist – producing as much as possible without taking costs into account. The idea of unlimited progress, anthropocentrism, the identification of development with economic growth, the identification of quality of life with disposable income and consumption, etc ... (Guterres, 2006).

According to Guterres, each family of farmers needs to become scientists of their own profession, learning from nature, from the behaviour of plants, animals and the environment, as well as seeking systematized knowledge in scientific studies that support and improve agroecology.

It is urgent that we have basic ecological agriculture schools to enable a new collective level of basic knowledge that gives minimum security to build another way of practicing agriculture, towards a firm and decisive transition to an agriculture free of chemicals and poisons and independent of large industry. An association of social movements with strategic partnerships – technical assistance, teaching and research institutions – should be established, seeking the formation of networks, regional and territorial forums and other forms of integration in which farmers and their families participate in the definition of lines of research, evaluation, validation and recommendation of appropriate technologies.[12]

However, technicians, mostly trained in a restricted view, do not have the ability to drive new ideas and continue, in practice, with an authoritarian stance, leaving an ever-greater distance between discourse and action, leading to a confused and contradictory scenario. As a result, although rural extension is considered an educational process, this is not revealed in its practice (Guterres, 2006).[13]

As we saw in the previous sections, agroecology is not a discipline, but a "transdisciplinary" approach and a new methodology of rural extension and technical assistance that discusses agrarian activity from an "ecological" perspective. It is a theoretical and methodological approach that intends to relate to producers in a new way, using several different scientific disciplines, However, there are differences between researchers-extensionists regarding the "north star" of agroecology and what forces will promote changes towards "rural development", "farm development", "sustainable development", and "socialism", the themes of our next section.

---

12   According to Friar Sérgio Gorgen: "the agricultural machinery and implements industry in Brazil was structured to serve the big guys. This is why they only manufacture tractors, harvesters, and large, sophisticated, heavy, and expensive implements. The small farmer needs to invest in light, simple, resistant, rustic, economic, and cheap mechanization" (Guterres, 2006).

13   See also Caporal (1991) and Caporal and Costabeber (2007).

7   The Heterogeneity of Agroecology: From Market Niches to Systemic Rupture

If Guterres draws attention to the role of public institutions, Altieri prefers a mix between public institutions and NGOs that have been working with farmers in recent years. For him, the urgent need to combat rural poverty and regenerate the resource base of small farms has stimulated several non-governmental organizations (NGOs) in developing countries to actively seek new strategies for the development and management of resources in agriculture. The work of NGOs is inspired by the belief that agricultural research and development must operate based on a "bottom-up" approach, using already available resources: the local population, its needs and aspirations, its agricultural knowledge, and indigenous natural resources. Strategies based on local participation, capacities, and resources are believed to increase productivity while conserving the resource base. The knowledge of farmers about the local environment, plants, soils, and ecological processes is of great importance in this new agroecological paradigm (Altieri, 2012, p. 41).

Some NGOs involved in Rural Development Programs (RDPs) have demonstrated a unique ability to understand the specific and differentiated nature of small production, promoting successful experiences in the generation and "transfer" of peasant technologies. A key element has been the development of new agricultural methods based on agroecological principles that resemble the peasant production process. For Altieri, this approach differs from that of the Green Revolution not only technically, by reinforcing the use of technologies with low input use, but also by socioeconomic criteria, with regard to affected cultures, beneficiaries, research needs, and local participation (Altieri, 2012, p. 41–42).

In his work and during his interview with the TV show Roda Viva (TV Cultura), in 2004, Altieri pointed out that herbicides and chemical fertilizers are relatively cheap due to government subsidies, which also applies to the final price of Green Revolution foods, as opposed to the apparent high price of agroecological products. Looking only at the tip of the iceberg, we believe that agroecological products are "expensive".

Altieri believes that the search for self-sustaining agricultural systems, with low use of external, diversified, and energy-efficient inputs, is the greatest concern of researchers, farmers and policymakers around the world. International agreements, such as the General Agreement on Tariffs and Trade (Gatt- now WTO), should continue to reduce or eliminate trade barriers and eliminate production subsidies.

In view of the institutional reforms, both Altieri and other researchers believe that decisions about public funds for research should explicitly take into account the environmental costs and benefits of the proposed research. The objective, pace, and direction of research in agriculture are key determinants for the level of adoption of agricultural technologies and productivity growth. Agricultural research has invariably been driven by relative prices or by the scarcity of land, labour, capital, and other production factors (Altieri, 2012).

According to Jalcione Almeida, Altieri's book *"Agroecology: the productive dynamics of sustainable agriculture"* is undoubtedly a powerful instrument for visualizing and making agroecology feasible as an area of knowledge and as a productive practice. Almeida believes that agroecology can put new forms of production and social organization on the agenda and contribute to a project that goes beyond the field of contestation and pure and simple opposition to technocracy, productivism, and inadequate agricultural policies (Almeida, 2012). Altieri does not have an agroecology project on the margins of capital agriculture, nor does it restrict the debate on the "greening" of modern or conventional agriculture, but rather as a form of agriculture poised as a true global technical-scientific alternative, as a renewal of the social and the technical-productive system, which can be a source of important cultural changes (Almeida, 2012).

Altieri's central argument is that new sustainable agroecosystems cannot be implemented without a change in the socioeconomic determinants that govern what is produced, how it is produced, and for whom it is produced. To be effective, development strategies must incorporate not only technological dimensions, but also social and economic issues. Only policies and actions based on such a strategy can cope with the structural and socioeconomic factors that determine the agricultural-environmental crisis and rural misery that still exist in the developing world (Altieri, 2012, p. 21).

Thus, for this researcher, the emergence of agroecology as a new and dynamic science represents a huge leap in the "right" direction. Agroecology provides the basic ecological principles for the study and treatment of ecosystems that are both productive and preserve natural resources, and that are "culturally sensitive, socially fair, and economically viable".

Also, according to him, agroecology considers the whole rather than just the parts and advocates minimum dependence on agrochemical and external energy inputs. Its principles are the preservation and expansion of biodiversity. At the same time, it provides a methodological structure of work, based on ethnoscience, which has cultural diversity and "respect" for popular knowledge as principles.

For Altieri, stable production can only happen in the context of a social organization that protects the integrity of natural resources and stimulates the harmonious interaction between humans, the agroecosystem, and the environment. Agroecology provides the necessary methodological tools for community participation to become the generating force for the objectives and activities of development projects. It is intended, therefore, that the peasants become the architects and actors of their own development (Chambers, 1983).

According to Caporal and Costabeber, the agroecological current suggests the massification of management processes and the design of sustainable agroecosystems, in a perspective of systemic and multidimensional analysis. Other currents, in turn, are guided mainly by the search for "niche markets, focusing their attention on the replacement of synthetic chemical inputs with organic or ecological inputs".

While the agroecological current defends an agriculture that is justified by its intrinsic merits, by always incorporating the idea of social justice and environmental protection, regardless of the commercial label of the product it generates or the market niche it will conquer, others propose an "ecological agriculture", which is exclusively guided by the market and by the expectation of economic gains that can be achieved in a given historical period. This model does not guarantee its sustainability in the medium and long term because, at the theoretical limit, a global ecological agriculture would not present a price difference due to the ecological or organic characteristic of its products (Caporal and Costabeber, 2002).

Despite its length, the passage below is illuminating of the problem mentioned here:

> While the agroecological current supports the need to build processes of rural development and sustainable agriculture that take into account the search for a balance between the six dimensions of sustainability, other currents, being guided mainly by the expectation of individual economic gains, end up minimizing certain ethical and socio-environmental commitments. From the perspective of an ecological agriculture devoid of these commitments, we can even assume that there a large-scale organic monoculture, based on poorly paid and treated wage labour might come to exist. This ecological monoculture may even meet the desires and whims of a consumer informed about the benefits of consuming "clean", "organic" agricultural products, free from contaminating residues. However, the degree of information or knowledge of this consumer may not allow them to identify or have knowledge about the social conditions in which the so-called organic product was or is being produced; perhaps

> they do not even want to know. In this case, within the theoretical limit and under the ethical consideration mentioned above, no product will truly be "ecological" if its production is being carried out at the expense of labour being exploited. Or, when the non-use of certain inputs (to meet market conventions) is being "compensated" by new forms of soil depletion or degradation of natural resources.
>
> CAPORAL and COSTABEBER, 2002, p. 80–81

According to Costa Neto (1999), we are facing the following crossroads: business ecological agriculture or ecological market agriculture (mercantile technological model) against family and settlement ecological agriculture (trending towards a socio-environmental model). Jalcione Almeida posits that agroecology relies on the potential use of social diversity and agricultural systems, especially those that actors recognize as closest to peasant and indigenous models.

> At the same time that they emerge and try to affirm new notions, agroecological actions and actors aim to put into practice a new type of collective movement, which will seek to get out of the more or less reclusive forms that most movements contesting social domination as a whole tend to fall into. But such a displacement of objectives, even if still of a strategic order and in an embryonic state, could not happen without great risks. Once again, the current condition of marginalization and exclusion of certain social groups and the urgent need to get results in the plan of social reproduction are factors that play against the capacity to solidify these new ideas, at least in the short and medium term.
>
> ALMEIDA, 2012, p. 241

Agroecology is not yet an organized social action against the power of its opponents, those who really have the reins of which type agricultural "development" takes place. A certain line of sustainable development wants everything to remain as it is, incorporating ecological demands without changing the substance of the capitalist mode of production: the exploitation of labour by capital. It would be a kind of "green capitalism" or "ecological capitalism".

In Novaes (2011a), we saw that the demands of social movements can point to an anti-capital society, based on decommodification, that is, on the production of non-poisoned food in agroecological research centres, or on what Mészáros called "positive transcendence of the alienation of work". But at the same time, they can be incorporated passively into the capitalist state, maintaining the pillars of domination and putting a halt to anti-capital struggles.

As we pointed out in this chapter, agroecology is heterogeneous. Theories of sustainable development are also extremely heterogeneous, as we'll point out in the next chapter.

CHAPTER 3

"Sustainable Development", Agroecology and Ecosocialism

1 Introduction

> The future cannot be a continuation of the past, and there are signs, both externally and, as it were, internally, that we have reached a point of historical crisis. The forces generated by the techno-scientific economy are now large enough to destroy the environment, that is, the material foundations of human life.
>
> ERIC HOBSBAWM, 1996

It seems the material foundations of human life on earth have brought us to a point of historical crisis. The Intergovernmental Panel on Climate Change (IPCC) report of August 2021 scientifically demonstrated that humanity is at high risk of facing serious environmental problems in the coming decades if current production and consumption patterns are maintained.

Research has demonstrated an increase taking place in the planet's temperature, especially in some regions, as well as intensification of hurricanes, unexpected frosts in some regions, water crises, desertification, and many other extreme weather events. António Guterres, Secretary-General of the United Nations, stated that this document spells "code red for humanity", with irrefutable evidence: gas emissions from the burning of fossil fuels and deforestation are suffocating the planet and putting billions of people at risk. Guterres also said that the report "must be a death sentence for fossil fuels before they destroy the planet". The head of the UN called for immediate action, including deep cuts in pollutant emissions, otherwise it will not be possible to limit global temperature warming to 1.5 °C.

This IPCC report was released in a very complicated year, as we are experiencing the coronavirus pandemic, aggravated in the Brazilian case due to Bolsonaro's criminal mismanagement of the pandemic. David Beasley, the executive director of the United Nation's food assistance agency, the World Food Program (WFP), said that the new coronavirus pandemic is causing widespread hunger "of biblical proportions" around the world. Beasley urged rulers to act before hundreds of millions starve in a short period of time. "We're not talking about people going to sleep hungry. We're talking about extreme conditions, an emergency situation. People are literally on the brink of starvation. If we don't get food for people, people will die", the director told *The Guardian*. It seems that governments' responses to the pandemic have been far askew of humanitarian needs.

Authors such as Eric Hobsbawm (1996), István Mészáros (2002) and Francois Chesnais and Claude Serfati (2003) state that capitalism can no longer be characterized by the extraordinary "development of productive forces", but of destructive forces, which are leading to processes destroying the conditions to sustain life on earth. Peasants and family farmers have faced many challenges. Various strategies have been developed by consumers and producers themselves to sustain ecological production. Both in Brazil and in Europe, the so-called Community-Supported Agriculture (CSA) model and the initiatives of producers to sell their products in stores in the city have been hugely successful.

Agroecology has been very promising regarding the production and consumption of healthy foods. The production of food without pesticides and synthetic fertilizers, using native seeds and with low use of tractors and agricultural implements, has found a promising market for the intermediate layers of society and some parts of the population that have enough income or greater ecological awareness.

But we must remember that much of social theory, including Marxist theory, was delighted with the fruits of the four industrial revolutions. New products, new processes, were invented and were considered "the good side of capitalism". Electricity, televisions, computers, cell phones, cars, airplanes are considered fruits of "technical progress" that allowed humanity, or to be more precise, a small portion of humanity, to live in better conditions. But it was in the 1970s that the first UN conferences to warn that spaceship Earth would not support the patterns of production and consumption stimulated by capitalism took place, and these warnings were based on scientific findings.

In 1962 Rachel Carlson published her book "Silent Spring" which issued clear warnings about destructive food production. Currently, in Brazil, the theories of Ana Primavesi (1920–2018), an Austrian who took part in creating the Federal University of Santa Maria in southern Brazil, are gaining strength. Her theories were set out in the book "Sustainable Agriculture" (Primavesi, 1986). José Antonio Lutzenberger (1926–2002) – an engineer who initially sold pesticides, changed his views and started to defend sustainable food production without the use of pesticides.

Luiz Carlos Pinheiro Machado (1929–2020) gave numerous lectures, courses, and field activities to build agroecological conversion processes, using the technique known as PRV (Voisin Rational Grazing). He was an international consultant for the promotion of sustainable agriculture. We should also remember that Chico Mendes (1944–1988) gained international relevance by denouncing the destruction of the Amazon, as part of the military's policy of "advancing the agricultural frontier" (1964–1985). In 1992, Rio 92 was held in Brazil, as part of the efforts for "sustainable development" and in 2012 Rio+20, which was nicknamed Rio-20.

When we finished writing this chapter (August 2021), a large demonstration took place in Brasilia in defence of the demarcation of indigenous territories. The 1988 Constitution assures the possession of the communal lands of indigenous peoples and Quilombolas. However, since then, there has been no effort by the Brazilian State to demarcate these lands, leaving indigenous people and Quilombolas in a totally unstable situation. With the advance of Bolsonaro's supporters, who are very interested in extracting ores and expanding animal agriculture in these lands, traditional populations are once again in danger.

Having contextualised the environmental problems that humanity has been experiencing, in the following pages we will try to highlight some of the contributions of sustainable development theories, but above all their limits.

2 "Sustainable Development" and Its Limits

Particularly in the 1980s and 1990s, theories of sustainable development gained strength, largely due to UN reports on environmental issues. We must point out that in Brazil, theories of sustainable development have undergone curious adaptations. A "new dictionary" emerges, generally created by economists, with words such as "sustainable growth", "sustainable agribusiness", "sustainable cities", "social and environmental responsibility", not to mention a huge business opportunity in the "green economy".

The 3Rs: Reduce, Reuse and Recycle, somehow entered the agenda of large companies, schools, governments, the State, etc. However, environmental issues within the political and theoretical frameworks of sustainable development can only address the issue in a rather epithelial way, which does not reach the root of the problems.

It is also necessary to remember that, despite symbolic advances in the environmental agenda, Brazil remains a well-oiled machine in the production of inequality. If we take stock of the New Republic (1986–2016) period, the concentration of income remains high, favelas continue to exist, underemployment still soars, and almost half of the population lives without basic sanitation and is under threat of food insecurity. Mining companies produced two major environmental-humanitarian crimes in the 2010s. Sugar and ethanol production is extremely destructive. How can there be 'sustainable development' in a country with so many social inequalities and environmental crimes?

One of the most important intellectuals in the theory of sustainable development is Ignacy Sachs (1927–2023). Sachs was one of the most renowned socio-economists or "eco-economists". A Pole who was forced to migrate to Brazil, he was an advisor to the UN and a central figure in the theories that culminated in the concept of sustainable development. Sachs was director of the School of Higher Studies in France and his theories had great international reach. He is cautious when assessing "boundless technological optimism" (Sachs, 1986, p. 32) and the possibility of appropriation of the productive forces engendered in capitalism by workers.

This social thinker believes that most of the technologies already available are not used to solve social problems because of political factors, especially the dominance of the ruling class, which prevents their use (Sachs, 1986). In his words:

> Paul Streeten is right to say that obstacles to development relate much more to human behaviour, social institutions and structures of political power than to the lack of factors of production and their correct allocation. And [Gunnar] Myrdal, as a good institutionalist, insists on the importance of what economists call "non-economic" factors in development. Keynes' great contribution to the debate on development was to teach economists in the Third World the priority of the political over the economic.
> SACHS, 1986, p. 103

In the book *Strategies for transition to the 21st century*, Sachs (1993) complements this idea by stating that:

> The essential nature of this obstacle [placing 1.5 billion people above the poverty line] is political and institutional, often related to inequality in land tenure, the lack of adequate agrarian reform programs, the privatization of common goods, the marginalization of forest peoples, or the predatory exploitation of natural resources, aiming at maximum profits in the minimum amount of time.
> SACHS, 1993, p. 27

To take just two examples, the introduction in Third World countries of "efficient and already known energy use techniques" would allow the South to achieve current Western standards of comfort with a tiny increase in *per capita* consumption. As long as there is political will, Sachs believes that a multiplicity of suitable technologies could be employed to reduce carbon emissions (Sachs, 1993, p. 36). Sachs is right to say that barriers to human emancipation are much more political than technological. However, if we have interpreted Sachs' works correctly, we can say that he somehow underestimates the productive obstacles that will exist in an eventual deepening of the premises necessary for the achievement of "ecodevelopment".

By theorizing development in its multidimensionality, Sachs (1993) laid out the construction of a society in which production covers the entire spectrum of material and immaterial needs and growth is subordinated to the logic of human needs. He believes that barriers to human emancipation are much more political than technological, but that there should be a change of research route in this area – especially in Third World countries – to develop appropriate technologies (Sachs, 1986).

He proposes several public policy challenges for science and technology, but does not understand them in isolation from other important actions. Among the macro-social policies necessary for "Ecodevelopment", the following stand out: the change in the lifestyle of human beings (remodelling of northern standards and endogenous and non-mimetic development for southern countries), as well as the need for the emergence of a longer time horizon, which economists are not used to. He defends the time horizon of ecology because he believes it will be the only one that enables "synchronic and diachronic solidarity" and "true development", which will mean the growth of wealth in harmony with saving natural resources and human development (Sachs, 1986).

Among the most specific challenges for S&T public policies, the author highlights "the commitment of all human ingenuity to value the potential resources of each ecosystem through appropriate techniques" (Sachs, 1993, p. 183). It is in this sense that Sachs poses three "obstacles" to be faced to achieve a "technological shift": 1) the need to adapt technologies to different ecological, cultural and socioeconomic contexts, instead of forcing the transfer of technologies only because they are available; 2) the contradiction between the research priorities established by the market or the military and the priorities indicated by a comprehensive analysis of social needs; 3) the growing distance between the great power of modern technologies and the already outdated systems of political and social control over them.

Investments and research in waste recycling, energy and water conservation, and increasing the life span of machinery and equipment will bring about successful solutions for job creation and environmental problem solving. Instead of biotechnologies, Sachs proposes the development of natural biopesticides and bioinsecticides (Sachs, 1993). He also makes a special mention of research and experimentation aimed at combining traditional and cutting-edge technologies, making them accessible to small producers and focusing on saving soil and water in the production of cereals, and using cereals rationally for livestock.

With his suggestions of changes in the way we research, Sachs highlights the extreme need to break the isolation of the sciences and make them enter dialogues with each other. For him, the segmentation of disciplines and specialization prevent a view of the whole and the complexity of the situations we are facing (Sachs, 1986). The generation and dissemination of new agricultural techniques – environmentally viable, economically efficient and adapted to the diverse needs of small producers around the world – will require "considerable effort" for many years (Sachs, 1993, p. 35).

Besides the changes already mentioned, Sachs also reinforces the idea that universities cannot continue to be "ivory towers" or "diploma factories". Rather, they should be a primary resource for local development (Sachs, 1993, p. 39). In our view, Ignacy Sachs makes a partial and incomplete critique of the destructive role of transnational corporations. His proposal for a socially fair, economically responsible and environmentally sustainable development, even if it has some positive aspects, fails to point to ecosocialism. Ignacy Sachs believes that there should be a change in the path of technological research– mainly in Third World countries – to develop appropriate technologies (Sachs, 1986 and 1993).

Just as an example, David Dickson (1980), in the book *Alternative Technology*, argues that contemporary problems associated with technology come not only from the uses for which it is employed, but also from its very nature. For him, technology fulfils a dual function: at the material level, it maintains and promotes the interests of dominant social groups in the society in which it was developed; at the symbolic level, it supports and propagates the legitimizing ideology of this society, its interpretation of the world and the position it occupies. Moreover, if one day the working class tries to appropriate the productive forces and make better use of them, it is likely that a significant modification of the inherited science and technology will be necessary.

Chesnais and Serfati (2003) point out that Marx already warned in *German Ideology* that a stage is reached in which, within the framework of existing relations, productive forces and means of circulation are born that can only become harmful. They are no longer productive forces, but destructive forces! For Mészáros (2002, p. 527), a concept that requires a fundamental reevaluation is the "productive advance of capital".

According to Chesnais and Serfati (2003), science, technology and ways of cultivating and manufacturing, or to put it another way, the forms of relations with nature, would be for socialism at the same time an inheritance and a springboard. They would first constitute an inheritance that socialism could accept after carrying out an inventory, at first without much detail. Then, it would be a springboard from which humanity could advance without having to perform more than route inflections and without having to manage immense damage when trying to reverse, at least partially, its consequences (Chesnais; Serfati, 2003, p. 46).

For them, technology and science were shaped by the objectives of social domination and profit, the mechanisms that guide the selection of science and technique (Chesnais; Serfati, 2003, p. 59). Behind the "autonomy of research that financial capital does not tolerate, even as a myth", there have always been powerful objective mechanisms, such as financing and ways of rewarding success, and subjective mechanisms, like the internalization of the values of bourgeois society that guided it according to the impulses of accumulation and the hierarchy of capitalism's objectives (Chesnais; Serfati, 2003, pp. 60–61). We could state that the hegemonic view does not perceive the social relations contained in technology and other productive forces, letting a perception of research autonomy and neutrality prevail.

Mészáros (2004) states:

> one of the most resistant illusions in relation to the natural sciences refers to its alleged "objectivity" and "neutrality", which are attributed to them by virtue of their experimental and instrumental character, in contrast to the socially more involved and committed character of the "human sciences". However, a more careful examination shows that so-called objectivity and neutrality are only legends, because, in reality, what happens is quite the opposite.
> MÉSZÁROS, 2004, p. 283 – emphasis in the original

Science is not a sovereign agent, materially and politically self-sufficient, says Mészáros. It is "inseparable, subordinate or 'linked' to the voice that dominates the present: the military-industrial complex business community" (Mészáros, 2004, p. 283 – emphasis in original).

Commenting on the opinion of Austin, who claimed that the great scientists arrived at their discoveries by "wandering to and fro with their instruments" and "stumbling upon something really important, rather than, one fine day deciding: let's tackle some problem", Mészáros (2004, p. 278) refutes the idea of independence of scientists by quoting Einstein, who, in his book On peace, stated

> if I were a boy again and had to decide how to make a living, I wouldn't try to become a scientist, an academic or a teacher. I would rather be a plumber or a peddler, hoping to find that modest degree of independence possible under the present circumstances.

Regarding the debate on the neutrality and non-neutrality of technology, for David Noble (1977), capitalist science & technology, far from allowing the historical emancipation of the working class, is an instrument of capital to strengthen its system of domination. Furthermore, the productive forces engendered in the sociometabolic system of capital would inhibit its reappropriation by others, highlighting the need for a radical restructuring in this field to debates on a socialist transition (Mészáros, 2002; Feenberg, 2002).

According to Mészáros (2002), we could make an analogy with the need to restructure the productive forces without going back to the Middle Ages, as in the case of Goethe's father's house. For Mészáros, the restructuring of productive forces and relations of production must encompass all aspects of the interrelationship between capital, labour and the state – and it is conceivable only as a form of transitional restructuring in the power of inherited and progressively changeable material mediations. Here, we could make a comparison:

> As in the case of Goethe's father (even if for very different reasons), it is not possible to put down the existing building and erect another with completely different foundations in its place. Life must continue in the house throughout the course of reconstruction, "removing one floor after another with completely different foundations in its place". Life must continue in the braced house throughout the course of the reconstruction, "removing one floor after another from the bottom up, inserting the new structure, so that in the end nothing must be left of the old house". In fact, the task is even more difficult than this. The deteriorating wooden structure of the building must also be replaced in the course of extracting humanity from the dangerous structural frame of the capital system.
> MÉSZÁROS, 2002, p. 599 – parentheses in the original

For those who believe that the productive forces incorporate the values of the society that generated it, not being guided by strictly technical criteria and that, in the capitalist case, do not contemplate self-management by the associated producers, it is essential to believe in history, a history that is essentially open, and that involves setbacks. Those who believe that there is already a predetermined path, with the succession of less and less oppressive modes of production until communism see history teleologically and ignore the role of the class struggle in it.

In this sense, Mészáros (2002, p. 527) states that the "productive advances of capital" are a concept that requires fundamental re-evaluation and that the productive forces generated in capitalism must be radically restructured. Therefore, they cannot be considered "the good side of capitalism".

John Belamy Foster (2005) believes that Marxism ignored or underestimated environmental issues in the 20th century. Marxist theories developed in the twentieth century did not give due attention to the destructive role of commodity-producing societies. To make matters worse, the division between the natural sciences and the humanities, which also existed in Marxism in some way, contributed to neglecting the observation of living conditions on planet Earth. The green parties that emerged especially in the 1970s failed to outline programs that minimally attacked the pillars of the sociometabolism of capital: alienated labour, private property and the State form of domination (Mészáros, 2002).

In Brazil, practically all of them are placed in the pro-capital, not anti-capital, field. They advocate improvements and refinements in the capitalist mode of production, but evidently fail to articulate a comprehensive program and actions in view of a socialist transition. The green party became practically a party for rent and was part of all recent governments, from Collor to Bolsonaro.

In our view, the approach of the theorists behind sustainable development is incomplete regarding the neutrality of science and technology and insufficient regarding to the serious problems that humanity causes. In one way or another, they do not overcome what we in Brazil call eco-capitalism. This seems to be the contribution of some aspects of agroecology to the critique of sustainable development and the construction of a transition theory based on ecosocialism.

3   Technological Dependence and Neo-colonial Reversal: Effects on Commodity Exports and Brazil's Role in the International Division of Labour

In order to observe the neo-colonial reversal in Brazil and the increase in technological dependence as a result of the productive restructuring of capitalism and its consequences for the export of commodities, it seems important to revisit dependency theory. According to Theotônio dos Santos (2000, p. 7):

> Dependency theory, which emerged in the second half of the 1960s, represented a critical effort to understand the limitations of a development that began in a historical period in which the world economy was already constituted under the hegemony of huge economic groups and powerful imperialist forces, even when a part of them went into crisis and opened up the opportunity for the process of decolonization.

For him, "the industrialization of the Brazilian economy took place in a peculiar way, in the absence of endogenous scientific and technological production and development, in the absence of formal or informal mechanisms to train the workforce for the new activities and in the absence of a significant or sufficient domestic market to sustain industrial growth" (Santos, 2000, p.8).

Authors such as Theotônio dos Santos (200) and Plinio de Arruda Sampaio Jr (2013) highlight the particularities of the industrialisation of underdeveloped countries and the enormous difficulties they have had or will have in achieving economic independence. However, despite highlighting the particularities of each dependent country, there is a certain consensus among Marxist thinkers in Brazil that the period of industrialisation of the dependent nations represented in some way a partial economic independence, which allowed the working class some gains from this new form of insertion into capitalism, via industrialisation.

For Sampaio Jr (2013), Latin American societies, exposed to the fury of globalisation and the discretion of rich countries, have been subjected to draconian mechanisms of neo-colonisation. Three processes are enough to characterise the perversity of "new dependence": a) the unequal diffusion of technical progress has increased the technological gap making backward economies lag behind, b) the transnationalisation of capitalism has reinforced financial dependence, which is evidenced by the structural nature of the imbalances in the balance of payments and c) the transformations in the pattern of capitalist development have intensified cultural dependence, compromising the basic premise of a national state: its existence as an entity endowed with its own "political will".

According to Campos (2023), "development" in the peripheral economies is increasingly induced by a more general dynamic of the imperialist diffusion of central economic structures, which in Brazil is absorbed through a path shaped by certain political decisions of the state. Countries that managed to partially industrialise in the period between 1930 and 1970 returned to the role of consumers of high-tech products and exporters of primary products in the phase of the globalisation of capital (1970 on).

It seems that the productive restructuring that has taken place in countries like Brazil since the 1980s has had consequences for our insertion into the international division of labour. In other words, the globalisation of capital has buried the possibility of our economic independence and put the export of primary products back into the core of our economic "destiny".

For Sampaio Jr (2023), Brazil's "neo-colonial reversal" took place especially from the 1980s onwards, with the opening up of the national market, violent processes of privatisation and financialization of the economy, among others. According to Otavio Ianni (2000, p.51):

> This is the irony of history: Brazil was born in the 16th century as a province of colonialism and enters the 21st century as a province of globalism. After a long and erratic history, through mercantilism, colonialism and imperialism, it enters globalism as a modest subsystem of the global economy. Despite outbreaks of nationalism and properly national achievements, as occurred mainly in the era of populism, i.e. an era which projected certain achievements for national capitalism, it enters the 21st century as a mere province of global capitalism; proving to be a case of perfect dependence.

Sampaio Jr. would probably agree with Otavio Ianni that Brazil has become a "province of global capitalism" and an exemplary case of "perfect dependence". (20023, p. 9):

> Converted into a mere "emerging market", the Brazilian economy became the target of veritable plundering operations by large international conglomerates interested in taking advantage of privatisations, mergers and acquisitions; using monopoly power to control entire segments of the national market and international trade; taking advantage of the weakness of states to extort huge tax and financial benefits; as well as exploiting comparative advantages arising from the control of strategic raw materials and the presence of cheap labour. The liberal-peripheral pattern of accumulation that emerged from this revitalised the primary-export economy, transforming Brazil into a kind of modern mega colonial outpost, based on monoculture and mineral extraction, geared towards the foreign market, organised around latifundia and large extractive companies, and based on the over-exploitation of labour and the overwhelming depredation of nature.

The neo-colonial reversal has profound impacts on the Brazilian economy, such as economic instability, an increase in the rate of labour exploitation, an increase in informality and a huge dependence on the export of primary products, as well as an enormous environmental impact.

Unlike the previous period, when there was partial industrialisation with positive consequences for economic growth, the strengthening of the labour market, the internalisation of production chains, etc., the production of commodities for the foreign market leaves Brazil in a totally subordinate position in the modern economy. With the 4th Industrial Revolution, Brazil is no longer able to insert itself differently into the international division of labour, thus becoming a modern neo-colony, with "perfect dependence" (Ianni, 2000) or a "modern colonial outpost" (Sampaio Jr, 2023).

In other words, the characteristics already highlighted by authors such as Florestan Fernandes (2020) and Theotônio dos Santos (2000), of technological dependence in the period (1900–1970) during our industrialisation, are now exacerbated, but with qualitatively different characteristics, as the country de-industrialises and becomes completely dependent on technologies generated in imperialist countries.

According to Plinio de Arruda Sampaio Jr (2023, p. 9):

> The revolutionary changes in the forces of production have intensified the hierarchical nature of the international division of labour because, in essence, the globalisation of production basically consists of the combination of high technology controlled by large corporations from the developed economies with the cheap labour force of the countries on the periphery of the world capitalist system. With the impossibility of a production goods industry anchored in the national economic space, the dream that it would be possible to achieve self-determination for capitalism has definitely become a chimera. National society has lost any possibility of putting limits on the vagaries of capital.

All consumer goods, from mobile phones to health equipment, are imported or produced using foreign technology. In rural areas, as we have seen and will see throughout this book, a large part of the inputs used are produced using foreign technology (pesticides, fertilisers, transgenic seeds, tractors, etc.). The Brazilian "industrialisation of agriculture" – in other words, the implementation of the "green revolution" agenda – is perhaps the world's best example of complete technological dependence.

From the perspective of World Systems Theory, Bunker and Ciccantel (2005) explore – in their book "Globalisation and the Race for Resources" – how five nations – Portugal, the Netherlands, Great Britain, the United States and Japan – achieved trade dominance by creating technologies, social and financial institutions and markets to improve their access to raw materials. Through an ecological and economic explanation of resource extraction and production, Bunker and Ciccantell (2005) reveal:

> globalization as the result of the progressive extension of systematically integrated material processes across cumulatively greater space. Drawing from extensive historical research into how economic and environmental dynamics interacted in the extraction of different materials in the Amazon, especially in the development of the iron mine of Carajás, the authors also illustrate the profound connection between global dominance and control of natural resources.[1]

4 The Contributions of Michael Lowy and István Mészáros to the Ecosocialist Transition

Marxist intellectuals have been increasingly concerned about environmental crimes and disasters. We highlight, among others, the studies of John Bellamy Foster (2005), with the book "Ecology in Marx"; István Mészáros (2002), in "Beyond Capital"; Joel Kovel and Michael Löwy (2003) with the "International Ecosocialist Manifesto"; Elmar Altvater (2007), "Is there an Ecological Marxism?"; Michael Löwy (2003) "Ecology and socialism"; and, by the same author, Löwy (2018), "Ecological message to comrade Marx".

In the field of Brazilian Marxist agroecology, it is possible to highlight the studies of Thelmely Torres Rego (2016), "Training in agroecology"; Dominique Guhur (2015), "Environmental issues and agroecology"; Wilon Mazalla Neto (2013), "Agroecology and Social Movements"; Henrique Novaes, Diogo Mazin

---

1 For more on the similarities and differences between Dependency Theory and World-System Theory, whose main author is Immanuel Wallerstein, see Dos Santos (2000 and Wallerstein (2004). World System Theory "studies the emergence, development and disintegration of historical social systems, researched using the comparative method, with the aim of arriving at generalisations about interdependencies between the system's components and principles of variation between systemic conditions in different spaces and times. Historical social systems are sets of structures (or entities), both systemic and historical, whose coexistence and succession represent the very content of the social world. These systems are seen as the most appropriate 'unit of analysis' for the study of social life and thus occupy the analytical place traditionally filled by 'society' and the 'state' as the entities in which social life takes place (Wallerstein, 2004)".

and Lais Santos (2015, organizers), "Agrarian issues, cooperation and agroecology"; Henrique Novaes (2017), "The world of associated labour and embryos of education beyond capital"; Sevilla Guzman and Molina (2013), "On the evolution of the concept of peasantry".

Outside of the Marxist sphere, but in dialogue with it, we can highlight the studies of Machado and Machado Filho (2013), "The dialectics of agroecology"; Ignacy Sachs (1986), "Spaces, times and strategies of development"; Ana Primavesi (1986), "Sustainable agriculture"; Petersen, Tardin and Marochi (2002), "Traditional (agri)culture and agroecological innovation"; Jan Ploeg (2008), "Peasants and food empires"; Jean Ziegler (2013), Mass destruction etc., which all contain fundamental contributions to "environmental" struggles.

Luiz Marques (2015) presents a rigorous study and is one of the most important in recent times at the international level. Springer Publisher recently published a translation. The author sheds light on the rise of transnational corporations in the twentieth century and carries out a thorough study of environmental collapse in various "fields". Marques carries out a broad and exhaustive analysis of the destructive power of corporations that culminates in a theory of environmental collapse, based on a wide range of scientific data.

Authors such as István Mészáros (2002), Michel Lowy (2003) and John Belamy Foster (2005) have brought fundamental contributions to the critique of "sustainable development" and the rescue of environmental issues within the Marxist perspective, in view of a theory that points to an overcoming of capitalism, a "diagnosis" combined with radical alternatives.

Michael Lowy believes ecosocialism can be a radical alternative. In his words:

> Attempts at moderate solutions prove completely incapable of facing this catastrophic process. The so-called Kyoto Treaty falls far short, almost infinitely short, of what would be necessary, and yet the US government, managers of the world's worst offending country, the champion of planetary pollution, refuses to sign it. The Kyoto Treaty, in fact, proposes to solve the problem of greenhouse gas emissions through a so-called 'market for the right to pollute'. Companies that emit more CO_2 will buy emission allowances from others, which pollute less. This would be the 'solution' to the problem of the greenhouse effect! Obviously, solutions that accept the rules of the capitalist game, that adapt to the rules of the market, that accept the logic of infinite expansion of capital, are not solutions, and are incapable of facing the environmental crisis – and because of climate change, it has put the survival of the human species at stake.
> LOWY, 2013, p. 81

Lowy (2013) also notes that the United Nations Conference on Climate Change, held in Copenhagen in December 2009, was another clamorous example of the inability – or lack of interest – of capitalist powers to address the dramatic challenge of global warming. It also notes that Rio+20, which tried to impose the so-called "green economy" – that is, capitalism painted in another colour – and ended with vague statements, with no effective commitment to combat climate change.

We agree with Lowy that we need to think about

> radical alternatives, alternatives that go beyond our historical horizon, beyond capitalism, beyond the rules of capitalist accumulation and the logic of profit and merchandise. A radical alternative goes to the root of the problem, which is capitalism. This alternative is ecosocialism, a strategic proposal, which results from the convergence between ecological and socialist, Marxist reflections.

Michael Lowy (1938-...) takes stock of non-socialist, capitalist or reformist ecology and shows all its limits, as we have seen above. Basically – for him – the limits of these proposals is precisely not to call into question class struggle and the ownership of the means of production. Lowy is a Brazilian Marxist thinker based in France, where he works as research director at the *Centre National de la Recherche Scientifique* (CNRS). He has dedicated himself to the theoretical construction of ecosocialism, among other equally important themes.

Lowy (2013, p. 82) also criticizes the non-ecological socialism of the twentieth century, taking as an example the "Soviet Union, where the socialist perspective was quickly lost with the process of bureaucratization, and the result was a process of industrialization with tremendous destructive impact on the environment".

Michael Lowy points out that, today, there is a global international ecosocialist movement. On the occasion of the World Social Forum in Belém, Brazil, in January 2009, he mentioned as examples of concrete actions the publication of a declaration on climate change and, within Brazil, an ecosocialist network that also published a manifesto.

István Mészáros (1930–2017) was born in Hungary. His experience as a worker and student in "socialist" Hungary was decisive for his understanding of real socialism and, later, real capitalism. His social theory of transition embodies a radical critique of destructive production. We believe that the "spinal cord" of Mészáros' theory is its criticism of the alienation of labour and the role of self-management in its "positive transcendence". Mészáros is an heir and at the same time a critical disciple of his master, Gyorgy Lukács and, above all, a

Marxist philosopher and one of the theorists who played an important role in the re-foundation of Marxism in the second half of the twentieth century. His book "Beyond Capital – Towards a Theory of Transition" contributes fundamentally to thinking about environmental issues from the Marxist perspective.

According to Ricardo Antunes (2022), Mészáros affirmed that, based on the structural crisis of 1972, the system of capital aims at self-valorization, not in order to expand, and proves to be uncontrollable. The destruction to the environment, work, and humanity has taken us to the limit.

And he continues:

> In the uncontrollable expansionist system of capital, based on an economy in which the war-hungry industry is powerful, all that was missing was the pandemic. And the pandemic isn't an abberation of nature. It is elementary: the more global warming, melting ice sheets, the more viruses that were there, spread across the world. The more forest fires there are, the more you destroy the habitat the more these organisms spread. The pandemic is a consubstantiation of this tragedy, of the destructive consequences of capitalism, of the mining of minerals, and the production of pesticides. If anyone says that I am exaggerating, I am, however, much less than I should be. We know how many people we have lost in the world, we know sleepless nights that we spent when we had close relatives contaminated at a time in which there weren't any vaccines to name of. It is a destructive, out-of-control, pandemic, and warmongering capitalism. It is a lethal capitalism.
> ANTUNES, 2022

Mészáros's (2002) reflection on the socialist transition takes place within the scope of the proposal he formulates, of a global change that aims at the transcendence of the "sociometabolism of capital". His theory seeks the qualitatively higher demands of the new historical form, post-capital (and not post-capitalist) socialism, where human beings can develop their "rich individuality". It is worth mentioning that Mészáros (2002) uses the expression post-capital and not post-capitalist because, for example, while the Soviet experience, a post-capitalist society, "extinguished" private ownership of the means of production but was based on a form of bureaucratized control, a post-capital society will extinguish all determinations of commodity production, and therefore, the control of sociometabolism will be in the hands of labour.

In his introduction of Mészáros's book (2002), Ricardo Antunes (2002) observes that Mészáros sees capital and capitalism as distinct phenomena and the conceptual identification between them made all the revolutionary experiences lived in the 20th century, from the Russian Revolution to the most recent attempts at constituting socialist societies, prove incapable of overcoming the system he designates the social metabolism of capital. Capitalism would be one of the possible forms in which capital realizes itself, one of its historical variants.

Ricardo Antunes (2002) also observes that Mészáros defines the system of social metabolism of capital as being powerful and comprehensive, having its core formed by the tripod of capital, alienated labour and the State – three fundamental and interrelated dimensions of the system as it is materially constructed – and it's impossible to overcome capital without the elimination of the set of elements that comprises this system. With no limits to its expansion, capital's system of social metabolic reproduction proves to be uncontrollable and obviously destructive of living conditions on earth.

In general, Mészáros' theory revolves around the alienation of labour and the need to overcome it. For him,

> the alienation of humanity, in the fundamental sense of the term, means loss of control: its embodiment in an external force that confronts individuals as a hostile and potentially destructive power. When Marx analysed alienation in his 1844 manuscripts, he indicated its four main aspects: the alienation of human beings from nature; from their own productive activity; from their species, the human species; and from one another. And he emphatically stated that this is not a "fatality of nature", but a form of self-alienation.
>
> MÉSZÁROS, 2006, p.5

In other words, it is not the feat of an all-powerful, natural or metaphysical external force, but rather the result of a certain type of historical development, which can be positively altered by conscious intervention to advance the process of transcending the self-alienation of labour (Mészáros, 2002). For the purposes of this book, it is necessary to hark back to Mészáros' general criticism of the sociometabolism of capital and his particular criticism of the so-called "Green Revolution" (as we did throughout the chapters). In our understanding, Mészáros believes that twentieth-century Marxism underestimated the environmental issue and was delighted with the wonders of capitalist "technical progress".

Likewise, it underestimated the role of Workers' Councils in the resumption of control of work processes and work products. For Mészáros, self-management means the resumption of control of the work process, the product of labour, oneself, and human civilization (Mészáros, 2002).

For him, another form of worker participation must be exercised in the microcosm and social macrocosm within a project to build a communal economy in the twentieth century. The need for authentic participation, rotation and revocability of positions are vital principles of self-management. How then does Mészáros resume the debate of socialist planning in view of what we are calling ecosocialism? He states:

> Those who despise the very idea of planning, due to the Soviet implosion, are very wrong, because the sustainability of a global order of sociometabolic reproduction is inconceivable without an adequate system of planning, administered on the basis of a substantive democracy by freely associated producers.
> MÉSZÁROS, 2004

The Councils have a potential to mediate and emancipate by rationally solving the workers' vital existential problems, the daily concerns with housing and work, the major issues of social life according to their elementary class needs (Mészáros, 2002). However, this author makes some warnings, since the Workers' Councils should not be considered the panacea for all the problems of the Revolution. That being said, without some form of genuine self-administration, the difficulties and contradictions that post-revolutionary societies have to face will repeat endlessly, and may even bring the danger of a recurrence in the productive practices of the old order, even if under a different type of personal control (Mészáros, 2002, p. 457).

Mészáros also reminds us that, when it was spontaneously constituted, in the midst of the important structural crises of the countries involved, the Workers' Councils tried to assign themselves on more than one occasion in history, "precisely the role of possible self-administrator, along with the self-imposed responsibility – which is implicit in its assumed role and practically inseparable from it – to perform the gigantic long-term task of rebuilding the inherited social productive structure" (Mészáros, 2002, p. 457). The resumption of Mészáros' theory of councils seems important precisely because we do not believe that an alleged self-regulated market could lead humanity to an "environmentally sustainable" future.

After this brief presentation of ecosocialist theory, we can now deepen the debate on agriculture, agroecology, science and social movements. Marcos Oliveira and Hugh Lacey (2001) observe the impossibility of transplanting "reductionist science", a component of the productive forces, in the case of food production. If the defenders of GMOs consciously or unconsciously argue in favour of transnational corporations and production focused on capital accumulation, that is, seeing seeds as a commodity, for the defenders of agroecology, a seed cannot be analysed only as a seed, but we must also analyse the social relations embedded in it.

Regarding the relationship between the modes of production, technology and knowledge used in agriculture, we could mention the militant-researcher Vandana Shiva. In the presentation of this thinker's book, Oliveira and Lacey state: "Shiva is a radical critic of the dominant technological models in agriculture and the knowledge that informs them" (Oliveira; Lacey, 2001, p. 17). For these researchers, the four types of violence she denounced – violence against the supposed beneficiaries of knowledge (poor farmers and their families), intellectual property (monopoly of knowledge), pillage of knowledge and plunder of nature,

> are a result not of particular ways of using this knowledge, but of their own nature. Reductionist knowledge necessarily serves the interests of capital-intensive agriculture and even in favourable socioeconomic conditions cannot contribute to projects favourable to social justice.
> OLIVEIRA; LACEY, 2001, p. 17

Oliveira and Lacey (2001) evidently do not follow the path of the "dead end" because Science and Technology carry contradictions. However, the arguments they use to defend a dialectic committed to the construction of another knowledge would be beyond the scope of this book.

Herein lies one of the contributions of agroecology from the perspective of Latin American social movements. The criticism of transgenics, the use of pesticides, of producing export monocultures on large land properties, the criticism of production relations based on exploited and/or alienated labour and hierarchical relations between men and women, are created in a perspective that opposes the production of commodities, that is, defends the production of food, the appropriate use of natural resources primarily to feed the people, and not to feed pigs that will be consumed in China. Finally, the production of values of use and not of exchange, or new social relations of production and consumption, are principles of food sovereignty.

The decommodification of agriculture, the self-management of production, agrarian reform, ecological issues, and healthy food consumption are all part of the struggles of social movements. They are based on new relations of production in the settlements for the restructuring of the productive forces (against transgenics, pesticides, etc.) and point to a new mode of production, to be built by the struggles of social movements.

More than that, these causes walk hand-in-hand with practical struggles that involve demonstrations against transnational corporations, actions in the settlements seeking new forms of work (based on cooperativism, associations and the promotion of equality between men and women), as constituent parts of the agroecological transition in the settlements. In our understanding, agroecology can articulate the present struggle with the struggle for future society. In Mészáros's (2002) terms, the articulation between the immediate needs of social and long-term movements, which can lead to a society beyond capital.

We'll see in the next chapter that in Gonçalves's (2008) point of view, what mobilizes the MST is the denial of the existing agricultural development model in Brazil, highlighting the need for the preservation and reconstruction of peasant agriculture through agrarian reform, in addition to proposing forms of management and participation of the peasantry in cooperative and agroecological production systems.

To the Via Campesina – an organization which seeks to build bridges between social movements throughout the world – agroecology is conceived of in a more politicized perspective than simple "sustainable development". In this sense, we believe that the Via Campesina makes a fundamental contribution to the development of the ecosocialist theory. In its own words, it defends

> The right of peoples, communities, and countries to define their own policies on agriculture, work, fisheries, food and land that are environmentally friendly and socially, economically and culturally appropriate to their specific circumstances. This includes the right to feed and produce your food, which means that everyone has the right to healthy, rich and culturally appropriate food, as well as the resources of food production and the ability to support themselves and their societies.
> VIA CAMPESINA, 2002 apud RIBEIRO, 2013

We can conclude this chapter as follows: We believe that agroecology is heterogeneous, as we saw in the previous chapter. In turn, the theories of "sustainable development" are also heterogeneous, with different implications for analysing reality, but above all for the solutions to socio-environmental problems put forward by researchers.

We can also say that agroecology is an offshoot of "sustainable development" theories. It is the result of research, actions by social movements, etc. that directly or indirectly contribute to the advancement of development theories, with the particularity of looking more deeply into agrarian and environmental issues in rural areas.

There are positive aspects in the more general theory of "sustainable development" and in the more specific theory of agroecology, which we have tried to explain in this chapter and the previous one. However, in our view, there are limits in the theory of "sustainable development" and in the more conservative theories of agroecology that prevent them from theorising ecosocialism. This is precisely where the contributions of Michael Lowy and István Mészáros seem promising, as we have tried to explore in this book.

CHAPTER 4

"Green Revolution" in Brazil, Rural Extension and the Fight to Establish MST

1   Introduction

Agroecology, as a socioproductive model, became more consistent on the Movement of Landless Rural Workers' agenda (MST) during the development of its fourth National Convention held in the year 2000 in Brasilia. The convention's formation was characterized by its intense debate proceedings and reflections regarding the development of a grassroots agricultural project, as well as a critique of the *Green Revolution* model and of its opposition to agribusiness.

The critique of the *Green Revolution* and the opposition to agribusiness has been gaining notoriety within MST since the preparation and completion of its III National Convention held in 1995. Since its III Convention, MST began thinking over the fact that the conventional model of agriculture which establishes the *Green Revolution* as a foundation imposes a series of negative and contradictory consequences to agrarian reform. However, it was only in its IV Convention that MST shifted gears making a commitment to develop an alternative and grassroots program for the base, rooted in another technological model – agroecology.

The *Green Revolution* was a production model which has been instituted since the 1960s with the goal of *modernizing* Brazilian agriculture, considered, until that time, to be outdated. The model was implemented on the basis of a technological shift, the addition of artificial input, the intense usage of agrotoxins, tractors, agricultural machinery, and genetically modified seeds.

The institutionalization of credit and its rural expansion, particularly during the military dictator (1964–1985), contributed to the diffusion of the *Green Revolution* wave emanating from *aid* deals together with the United States government and philanthropic institutions, the majority of them also being from the United States. The diffusion of the *Green Revolution* agricultural model was so intense that MST wasn't immune to it, so much so that this model was more assertively established in the transition from the 90s to the 2000s, in a context in which agribusiness had also been taken up.

For MST, agroecology represented a point of divergence from agribusiness' interests and to the shift in the scientific paradigm in relation to the *Green*

Revolution wave. Upon adopting agribusiness, MST took on the role of amplifying the debate to the masses in regards to the scientific proposition and to the technological alternative for rural Brazil. Luiz Carlos Machado and Machado Filho (2014) affirmed that agroecology is a scientific and technological production model that aims to recover traditional knowledge hidden by the hegemonic agricultural model, critically incorporating and developing technical and scientific progress and utilizing and transforming natural resources which before were less degrading.

To promote agroecology in the camp and settlement territories for agrarian reform, MST, in the state of Paraná, determined that education and training of its activists was a fundamental factor. Among the adopted tactics of putting this concept in practice, we emphasize that the creation of formal and informal agroecology courses, the construction of Centers/Agroecology Schools[1] and the annually held Agroecology Excursion, were some of the tactics adopted to put this concept into practice.

2   The "Green Revolution" in Brazil

In Brazil, the *Green Revolution* introduced a structural framework to the camps that reorganized them while inserting mechanization, petrochemical input, optimized plants and seeds, and agribusiness enterprises. This process of organization and/or reorganization of the camps subjugated small rural farms to these rules and resolutions to what is now currently known as agribusiness (GONÇALVES, 2008).

The *Green Revolution* was accompanied with an intense scientific and technological shift, that aimed to increase agricultural productivity. Albeit, the *Green Revolution*, as a part of the structure of subordination of agriculture from modern centers, resulted in a process of substituting workers in the camps for machinery and ramping up its dependency on it, by means of using chemical material, originating from foreign industries (Medeiros, 2001 *apud* Guhur, 2010).

Particularly regarding the development of Brazilian capitalism, the modernization of agriculture established the *Green Revolution* as a technological model to carry out the process of subcapitalization of Brazilian agriculture. Without enforcing modifications on the structure of land-ownership, such a

---

1 The Centers/MST Agroecology Schools in Paraná are: the Latin American School of Agroecology (ELAA), Center for Sustainable Development and Training in Agroecology (CEAGRO), José Gomes da Silva School (EJGS), and Milton Santos School (EMS).

process exacerbated Brazilian dependency, be it on a political-economic level or on the level of the artificial material used to boost agricultural production to an industrial level.

The antagonism between capital and labor worsened with the *Green Revolution,* and the worker increasingly became more alienated from his labor production, to the extent that the camps were converted into agriculture business and started to work from within the capitalist industrial logic. Due to this agro-industrial characteristic, subordination of natural resources worsened in detriment to capital's productivity, as well as to the subjugation of the worker, who went from being a permanent employee to a temporary employee (SILVA, 1981).

> The essence of the *Green Revolution*, today, controlled explicitly by financial capital, which a small group of multinational companies govern that own patents for seeds, the production of fertilizers, and agrotoxins, and is changing the environment and implanting menacing monocultures, incorporating large energy contingents, via "modern material" that, in turn, are controlled by financial capital that, thereby, carries out the reproduction of capital in a new economic segment, agribusiness or the agriculture industry.
> MACHADO; MACHADO FILHO, 2014, p. 54

The *Green Revolution* understands, within the agricultural backdrop of the XXI century, the fundamental part of agribusiness, which is characterized by the predominance of financial capital and its union of industrial capital with banking capital. Within this dynamic, Brazilian agriculture underwent a process of development with the influence of a limited sector of multinational corporations that take national land, dominate, and control the food production chains and the market of chemical fertilizers.

In light of this context, with the advancement of agribusiness there was a realignment of the organizations from the Brazilian property-owning class, such as: National Agriculture Council (CNA), which already has a history of advocacy and political engagement in the National Congress through the Ministry of Agriculture, Livestock, and Food Supply (MAPA), the Brazilian Association of Agribusiness (ABAG), National Union for the Sugarcane Industry (ÚNICA), Program of Studies of the Agro-industrial Business System (PENSA) in defense of its interests.

These interests can be characterized, for example, by the acceptance of the so-called *biotechnological revolution,* implanted by multinational companies which immorally applied intellectual property rights to genetically modified seeds, together with the patent laws, exclusively edited to protect piracy by the agribusiness corporations (Machado; Machado Filho, 2014, p. 60).

> Biotechnology and transgenics, in the manner in which they have been utilized in agricultural production are reductionist techniques that promote monocultures and produce severe genetic and laminar erosion. Without regard to the harmful effects that the consumption of its products cause to human health, they are practices that eliminate biological diversity, impeding the natural genetic improvement of the populations.
> MACHADO; MACHADO FILHO, 2014, p. 80

The patent system acts as an agent of genetic erosion of biodiversity and one more element causing the impoverishment of many populations and nations, to the extent that the privatization of transgenics and biotechnology criminalizes agricultural practices that do not pay the franchises for the use of the invented technology by the large corporations, which is quite often built on the basis of the appropriation of traditional knowledge (Shiva, 2001).

With patent laws and the control of seeds by multinational corporations, we can consider the Brazilian situation to be somewhat delicate and vulnerable, given that the interests of these businesses threaten the sovereignty of the country. Machado and Machado Filho (2014) exemplify the Kandir Law, as an example of this vulnerability, seeing that this law excludes multinational companies from paying taxes on financial transactions and facilitates the registration and the usage of poisons by the National Technical Commission of Biosecurity (CNTBIO), even when these same poisons have been outlawed in other countries. In this regard, agribusiness contributes to the contamination of natural resources and to the increase of the country's dependency on agricultural materials produced in other countries, revealing that the practice is unsustainable, such that Francisco Caporal shows us:

> Brazil consumed, in 2007, around 10.6 million tons of NPK, in other words, we were fourth among the countries with the highest worldwide consumption. This model established an outrageous dependency on the importation of NPKs for our agriculture, because to sustain this agrochemical agriculture the country imports 60% of the Nitrogen, 40% of the Phosphorus, and 90% of the Potassium that we utilize. On a whole, this means an external dependency on 66% of the NKP used in agriculture. And this dependency shows tendencies to grow. Nevertheless, there are studies demonstrating that the average productivity of some farms

don't respond anymore to the elevated doses of chemical fertilizers. On other farms, there has even been a decrease in productivity.
CAPORAL, 2011, p. 132

According to data from the National Fertilizers Dissemination Association (ANDA), in 2015, Brazil imported more than 21 million tons of intermediate fertilizers and delivered more than 30 million tons to the final consumer, among them being fertilizers both imported and produced in the country (ANDA, 2016).

In light of this, it is necessary to draw attention to the possible lack of minerals such as phosphorus and potassium. There is information saying that the world reserves of potassium, for example, amount to 16 million tons while that of phosphorus are estimated to be at 50 billion. With a basis in this data, there are studies that highlight that in 2025, the strategic minerals used for the production of fertilizers on an industrial scale could be depleted and will also be considered a factor for economic security for the nations. For that matter, in a country that depends on capitalism such as Brazil, problems with autonomy and self-determination regarding who to feed the country can worsen (Caporal, 2011).

Considering that approximately 65% of the fertilizers and 100% of the pesticides utilized in Brazilian agriculture are imported or produced in the country by multinational companies, one can identify the fragility and dependency of the country's agriculture on interests which are external to those on the national level. Because of this, Machado and Machado Filho (2014) conclude that Brazil, in addition to having its food sovereignty threatened, can also have its political sovereignty threatened.

Agribusiness, in addition to its large dependency on inputs and fertilizers, has its aggravating factor potentialized when the ecological question, biodiversity, and traditional knowledge of indigenous people are in debate. Monocultures have been responsible for the loss of biodiversity in all of the Brazilian biomes. For example, data from the International Conservation Organization – Brazil demonstrate that of the 204 million original hectares of the Brazilian Cerrado (savanna) 57% has already been destroyed, the annual rate of deforestation is alarming, reaching 1.5% or 3 million hectares per year (Caporal, 2011). In addition to the Cerrado, the Atlantic Forest, according to a study published by the Brazilian Institute of Geography and Statistics (IBGE), suffered from the loss of approximately 102, 938 hectares of its native forest coverage or two-thirds the size of the city of Sao Paulo, between the years 2005 and 2008.

In the case of the Brazilian Amazon, the situation is also bleak, Caporal (2011) highlights that progressive deforestation, takes into account the amount of clear-cutting or deforesting from November 2008 to January 2009, totaling 754 km², with the invasion of the illegal agricultural borders. The author still highlights that while the Amazons, deservedly, has been receiving the focus of attention, the other biomes have been highly impacted. Besides deforestation and consequently the loss of biodiversity in Brazilian biomes, agribusiness also generates the contamination of the soil and sources of water, be they in the streams of the riverbeds or in underground water.

An astounding fact points us to the loss of more or less 500 tons of soil, one hectare per year, in endangered areas such as the Pantanal (Brazilian wetlands), a consequence of industrial agriculture and conventional livestock farming that has been causing the decrease in organic materials, consequently causing desertification in these areas. Another maleficent consequence of this process is the silting up of the steams, rivers, lakes and reservoirs, in addition to contamination of groundwater by pesticides such as the Guarani Aquifer, one of the last fresh water reservoirs of the country and of South America (Caporal, 2011).

In addition to the ecological and genetic erosion, the scientific and technological model of agribusiness and the capitalization of small farms contributed to the espousal and alienation of the workers from the countryside and the forest areas, causing them to become dependent on the multinational producers of fertilizers and agrotoxins. It is important to mention that these pathways were part of a process, as Florestan Fernandes said, of modernization of the "outdated", or as Teotônio dos Santos stressed, of the abandonment of efforts towards a regional scientific and technological production, which, by extension, creates difficulty for the capital goods sector, and for endogenous and alternative scientific and technological production.

3   Importing the Model of Rural Extension

The service of Technical Assistance and Rural Extension (ATER) was officially institutionalized in Brazil in the 1970s, with the the political, economic and cultural introduction of the technological model called the *Green Revolution*, which aim to *modernize* agriculture. Despite ATER having been consolidated during the military dictatorship, it is important to observe that the period that preceded the regime implemented in 1964, therefore, before the political coup d'etat, already had its first contact with foreign organizations that provided the underpinning for ATER's institutionalization with a foundation in the technological model of the *Green Revolution*.

Oliveira (2013), analyzing ATER's history in Brazil, emphasizes its process of inception originating with bilateral agreements together with the United States of America (USA), initiated in 1945, by means of technical cooperation agreements. The two countries injected material and intellectual resources into the bilateral agreement, with the perspective of exchanging technology and training Brazilian technicians via exchange programs and training resources in the USA. So that we can understand this process, it is imperative to explain, at least, how the service of rural extension in the United States was built. The rural extension structure in the United States has a decentralized organization suitable for each state, and, in turn, for each city to organize the labor of the rural extension agent. In view of this, rural associations, in particular the Farm Bureau (a civil organization of farmers), establishes the contract rules between extensionists and the farmer, therefore, the service of rural extension depends on resources handed down by means of regulated contracts between extensionists and the rural associations in each municipality (Oliveira, 2013).

Considering the particularities of American legislation, the service of rural extension can present a differentiated standardization, in some states, for example, ATER could be developed by means of the organization of associations, where relationships between the extensionists and the rural farmer occur in a more or less horizontal manner (Oliveira, 2013). For Bechara:

> Rural associations lay down the rules of the contract between the Extensionist and the Farmer. Thus, the project's 'volunteer rural leaders', highly propagated into the extensionists' practices, was a mirror of the traditional leaders of the local associations that emerged from the civil associations in which they represented. This means that it was the community itself that recognized these leaders. However, the organization of the rural extension services in Brazil would be totally distinct, had it not been for the difference between the social structure of both countries. The Brazilian extensionists, invested in the supposedly in-depth knowledge, it would be the extension worker that select for himself, through his contact with the community, who would be suitable to become a valuable rural leader. The choice was not made by the volunteer organizations of civil society, but by agents outside of the rural communities who have the necessary 'knowledge' to select such leaders.
> BECHARA, 1954 *apud* OLIVEIRA, 2013, p. 30

Rural extension, inspired by the American model and imported by Brazil, contributed to the dissemination of socioeconomic diagnostics that characterized the Brazilian agrarian situation as *outdated*, which, among other things, reinforced the argument for the need to *modernize* via the acquisition of technology and the implementation of the *Green Revolution*.

Regarding this, Machado and Machado Filho (2014, p. 43) describe that the:

> The self-proclaimed 'conservative modernization' of agriculture was neither a 'revolution' and to a lesser extent, nor was it 'green'. The expression 'conservative modernization' still includes a contradiction: because modernization is antagonist to conservation. This was implemented with the usage of 'modern inputs', a euphemistic neologism to refer to seeds, soluble fertilizers from chemical synthesis and industrial pesticides, with highly subsidized credit, and the ludicrous idea of modifying the environment by replacing natural factors with modern inputs. Its implementation in Brazil was monitored by an intelligent preparation, which counts on the support of the MEC-USAID agreement, that, in addition to modifying the structure of the Brazilian university also fostered the training of hundreds of Brazilian technicians with post-graduate degrees in the United States of America. These technicians received scholarships that funded them while in the United States for four years and also allowed them to reasonably save money … Foreign debt funded it.

José Graziano da Silva (1981) ironically highlights that *modernization* was a "magic solution", that would have us believe that increasing agricultural productivity would increase revenue, and consequently, also jobs and salaries in the rural zones. Moreover, revenue remained low for the majority of farmers, as well as job opportunities were strictly reduced to a limited number of workers.

Octavio Ianni (2009) indicates that *modernization* occurred, in various aspects, by means of capitalism's intense and massive development within the rural areas' social organizations in a manner that:

> The agricultural industry and the city, that is to say, the social classes and the urban-industrial base, particularly, the industrial banking, and commercial bourgeoisie largely expropriated the rural classes, salaried workers (permanent and temporary), small farmers, share croppers, rural landowners/squatters, spouses of landowners, partners, leaseholders, and others. A large part of these workers end up supplying successive contingents of labor to the agriculture industry. In this sense, agriculture has largely served the industry, as a area that stores an important portion of the industrial reserve army. Not to forget that a portion of these reserve-workers of whom the industrial capital always counts on have been utilized in the "occupation", "colonization" or "expansion" of the

> internal "limit" of capitalism. In the Amazon forest and the Central-West, there are substantial contingents of rural workers involved in a unique process of primitive accumulation that continues to develop.
>
> IANNI, 2009, p. 161

In this light, the process of *modernization* of agriculture has turned it into a consumer market for industrial products, that ranges from household appliances to *mechanization* and *chemicalization* of the production processes, and the agricultural production model itself has been managed like an industry. Furthermore, there has been action on the part of the state which boosts the connections between industrial capital and agriculture, financing the infrastructure necessary for *modernization*, such as the construction of roads, fiscal incentives and credit, "in order to attract large-scale business or promote pre-existing companies" (Ianni, 2009, p. 164).

Caporal (1991), when analyzing agriculture's *modernization*, states that the process of imperialist intervention and amplified reproduction of capital in Latin-American agriculture, particularly by the U.S.A., is based on dependence and domination which has been practiced and intensified post-World War II.[2] According to Xavier (2008, p. 15), the concept of structural dependency, from the perspective of dominated societies, reveals that this "is the result in a dominated society, resulting from the imperialism that manifests in the dominating society".

> In the framework of dependent capitalism, Brazil is subordinated to the role of a producer of industrial goods and a consumer of capital goods in the international division of labor, a product of the juxtaposition between external mandates and internal determinations. 'The industrialization of the Brazilian economy' operated peculiarly in the absence of production

---

2 The industrialization process of the dependent economies entered the international division of labor in the capitalist world as a new way of realizing the expanded reproduction of capital: absorbing and guaranteeing the profitability of surplus capital in the hegemonic centers, producers of production goods, with producers of consumer goods. The imperialist pressure was only felt in Brazil from the 1950s onwards due to the start of the second phase of US capital exports, when the hegemony of these capitals in the international market ushered in the era of direct investments, as opposed to the portfolio investments that characterized European capitals. In the first phase, which lasted until about 1953, the penetration of US capital in Latin America was moderate, since Europe, which was rebuilding, absorbed most of the US aid. After 1955, when European reconstruction ended, the penetration of US capital turned massively towards the Latin American countries that were beginning their industrialization process (Xavier, 1990, p. 44).

> and scientific and technologically endogenous development, formal mechanisms or informal labor training mechanisms for new activities, and in the absence of a significant or sufficient internal market to sustain industrial growth. Hence the need to import technology and, to this end, capital; to import labor force, at least in the initial phase; and as for the foreign market, a trend will solidify with the exhaustion of the so-called 'import substitution model'.
> XAVIER, 2008, p. 18–19

The concept of structural dependence, which gained momentum in Latin America in the 1960s, "tried to explain the new characteristics of socioeconomic development of the region, initiated de facto in 1930–45" (Santos, 2000, p. 25). It was an environment of reorientation after the crisis of 1929, reorientation in the direction towards industrialization, characterized by the replacement of industrial products, imported by the central economic powers, with domestic production.

This stance generated industrial growth, that between the 50s and 60s intensified the contradictions between capital and labor, to the extent that it accentuated social inequalities. The Brazilian bourgeoisie discovered that the expansion of industrialization demanded an agrarian reform and other structural changes that could leverage the creation of a large domestic market and the establishment of an intellectual, scientific, and technological base capable of sustaining the alternative project; such changes fomented an ample political and ideological upheaval in the country, that threatened economic stability.

Within this dynamic, a political coup d'etat occurred in 1964, that among other things, closed the doors to national democratic progress and deepened dependent development, supported by international capital and in strategic consensus with the world power system. "What is good for the United States is good for Brazil – a strategy from general Juracy Magalhães, minister of Foreign Relations for the military regime, consolidated in this direction" (Santos, 2000, p. 34).

> Dependent economic modernization implicated cultural and institutional modernization which, like the economic modernization tended to take place within the limits necessary for the incorporation of the national economy together with world capitalist economy to which it was subordinated.
> XAVIER, 1990, p. 58

In this light, it was under the aegis of foreign *aid* programs, associated with technical aid programs, that the process of international cooperation occurred and the consolidation of dependence of the peripheral counties to those of the central countries. Finally, the technical and financial cooperation policies accomplished the political-ideological role of masking international capitalism's true interests, representing one more sphere of the circular relation of dependency, which prepared Latin American countries for the introduction of foreign capital, substantiated by the political-economical project constructed by the military dictator (Minto, 2006).

The anxiousness to overcome the *outdateness* in the countryside contributed to the establishment of dependent capitalism, and it was up to the state to guarantee instruments for the internalization of the technical progress, in order to create a path so that agriculture accomplishes its function of increasing productivity and creating a financial reserve through commodities. To reach these goals, Caporal (1991) affirms that both the Brazilian state as well as the American state significantly invested in material, technical, and human resources.

It is in this context that we highlight the presence of the American International Association (AIA), a philanthropic wing of the Rockefeller[3] group in Brazil, that working with the governor of Minas Gerais in the year 1948, founded Credit Assistance and Rural Assistance (ACAR). ACAR's objective was to establish "a technical and financial assistance program that enabled the intensification of farming production and the improvement of the economic and social conditions of rural life" (Caporal, 1991, p. 33). According to Peixoto (2008), ACAR was the civil entity, non-profit, that provided services for rural extension and the drew up technical projects to obtain credit from financial agents.

At the same time of ACAR's growth, the proposed bill n° 2.613, from September 23, 1955, provided for the foundation of Rural Social Services (SSR) within the Ministry of Agriculture, an example of the already existing Industrial Social Service (SESI). SSR was an autonomous body, with legal personality and its own capital, with a headquarters in the Federal District and jurisdiction in the entire national territory.

---

3 It is worth highlighting the figure of Nelson Rockefeller as one of the Brazilians' allies in the introduction of rural extension. However, as Karavaev (1987, p. 126) pointed out, his interests in Brazil were spread across various fields, where he had internal agents. In 1947 (the year the extension project in Santa Rita do Passa Quatro-SP was set up, under the sponsorship of the AIA), a commission was created by the government to regulate the oil issue. Braga, president of Gás-Esso, a Brazilian subsidiary of Rockefeller's Standard Oil Company" (Caporal, 1991).

SSR aimed to promote social campaigns in the countryside, such as: construction of trenches, medical assistance, courses, among other things. Its activities were conducted in the spirit of *aid* and social guardianship seeking to provide services in rural areas, aiming to improve the living conditions of the populations on the countryside (Oliveira, 2013).

Oliveira (2013) points out that, despite some progress, SSR demonstrated a huge limit in performance, a part of this limit is represented by the performance of the Brazilian Rural Confederation (CRB) and the National Society of Agriculture (SNA) employer organizations subordinated to the Ministry of Agriculture that opposed the consolidation of SSR. In 1956, the Brazilian Association of Credit and Rural Assistance (ABCAR) was founded, following a vertical model of orientation.

ABCAR was under the influence of foreign agricultural policy and of the regulation of the national employer class, in which rural assistance had the goal of maximizing agricultural production. Despite being created as a non-profit organization and coming from private civil law, it possessed a direct relationship with the Ministry of Agriculture, be it for the allocation of resources or for the maintenance and recruitment of technical personnel (Oliveira, 2013).

The institutionalization of ABCAR, according to Caporal (1991), was profoundly characterized by dependency and the implementation of the North-American model, to the extent that AIA and ETA (Technical Office Brazil – United States) were its founding members and maintainers, together with Bank of Brazil (Banco do Brazil), CRB and its affiliates. Subsequently, the Ministry of Agriculture, the Ministry of Education and Culture (MEC), SSR, Brazilian Institute of Coffee (IBC), and National Bank of Corporate Credit (BNCC) joined. Within this dynamic, American technicians integrated with ABCAR's Consultancy, and Brazilian technicians obtained training opportunities in the United States (Caporal, 1991, p. 38).

As a result of ABCAR's system, in 1974 the Brazilian Entity for Technical Assistance and Rural Extension (EMBRATER) was created, which takes on the state's activities, as its strong arm, along with the rural population, having the role of enhancing *modernization* and guiding the process of implementing the *Green Revolution* package.

> The creation of EMBRATER and, later, EMATER gave the state a new power to operate together with the rural environment, since as the ministers said in the Explanatory Memorandum n.º 08/74 which proposed the creation of the Brazilian Entity for Technical Assistance and Rural Extension (EMBRATER 1975:10) to Congress, which made having a "mechanism of flexible and powerful operation", a "strong and dynamic organization", a

> "fast instrument and efficient for the execution of integrated programs". necessary. Since then, state-owned companies' Technical Assistance and Rural Extension's activities satisfy greater interests, established in the government's plans, both on the federal level and state level.
>
> CAPORAL, 1991, p. 59

The establishment of EMBRATER occurred soon after the establishment of the Brazilian Agricultural Research Agency (EMBRAPA), both of them fulfilled the goal of advancing the direct action of the state in spreading the *Green Revolution*. In the same vein, the Ministry of Agriculture created, the National Commission for Agricultural Research and Technical Assistance and Rural Extension (COMPATER), by means of the Bill n° 74.154, of June 6, 1974.

> EMBRATER coordinated, in the country, the state's initiatives for the rural areas. This, to a certain degree, gives joint responsibility to both EMPRESA and SIBRATER – the Brazilian System of Technical Assistance and Rural Extension for the outcome of the implementation of the urban-industrial model of development, which made the countryside a stage for large social transformations, of large exclusion of workers and their families, and above all, of intense and continuous subordination to industrial, commercial and financial capital.
>
> CAPORAL, p. 35, 1991

COMPATER lasted for a short period of time, having been discontinued by the Bill n° 86.323, of 31 August 1981 which transferred its mandates to the National Secretary of Agricultural Production to the Ministry of Agriculture. Consequently, between the decades 1950 to 1970, the country accelerated the process of modernization of agriculture, from the model of rural development based on the transfer and diffusion of technological foreign packages (Peixoto, 2008).

So that modernization could be implemented, Leite (2001) points out the importance of credit incentives, carried out by the National System of Rural Credit (SNCR) created by the Law n° 4.829, of 05 November 1965 and regulated by the Bill n° 58.380, of 10 May 1966. The credit could be granted by the Central Bank of Brazil (BACEN), by Bank of Brazil (BB), by regional development banks, by state banks, by private banks, by savings banks, by credit societies, by co-ops and branches of ATER. Delgado (2001), reinforces that subsidized rural credit was one of the fundamental tools of Brazilian agricultural policy in the 1970s, made possible by the circumstances of high liquidity in the international credit market and in the domestic monetary system, allowing for the installation of an expansionist credit policy in the country.

Marine (1987) affirms that access to credit was essentially so that the *Green Revolution*'s technological package could be bought, and its selling points promised an increase in average productivity through enhanced seeds or high yields, while the utilization of these seeds were conditional and integral to the use of machines and chemical inputs. Due to these characteristics, the rural extension and credit policy model amplified the production of monocultures on a large scale, the monopoly of seeds by transnational companies, and the intense usage of agrochemical products coming from industry.

In economic and productive terms, the installation of the *Green Revolution* lead to increasing production costs due to the intense usage of chemical fertilizers, pesticides, and due to the deterioration of soil and water resources which made natural resources scarce.

> The size of big farms augmented substantially [...]. The rate of adopting new technology was directly related to the size of the property [...]. There was a quick adoption of biological technology, especially, chemical fertilizers; this process of adoption significantly increased operation costs [...]. There was a dramatic increase in the use of agricultural credit in recent years; all of these increases in the supply of credit was channeled through formal institutions of credit; [...]; real negative interest rates generally predominated and distorted the allocation of capital and credit; real negative interest rates also resulted in the substantial transfer of income to credit users.

The policy of subsidized rural credit adopted in Brazil between the period of 1960 and 1980 had the goal of compensating the agricultural sector for the negative effects from the trade, exchange rate, and fiscal policy and still carry out the *modernization* of Brazilian agriculture. The low-cost financing on the transaction was directed to huge land owners because they possessed the structure that guaranteed a better capacity to pay, a fact that, clearly, benefited the conservative pact and reinforced the concentration of large landowners.

> Small landowners were harmed by this policy because they did not have access to the banking system. In addition, the use of inputs (technological package) was linked to the acquisition of subsidized credit, cheapening credit (and at the same time making the labor force more expensive), causing a distortion in the allocation of resources, insofar as

> labor force (abundant) was dispensed, unable to benefit from the comparative advantages, causing underemployment and a rural exodus.
>
> GUIMARÃES, 1997, p. 123

This position accentuated the inequalities even more among, small, medium, and large farmers, contributing to the growing misery of the small farmers, to the extent that they needed to dissolve their means of production, given the difficulties or even impossibilities of maintaining and surviving even before the expansion of capitalism in the countryside. Many of the small farmers had to increase their work hours and that of their families, for some, this meant selling their labor force to others.

According to Pretto (2005), the possession of land was a requirement demanded for obtaining bank credit, the only condition by which land was given a market value due to financing that was passed on with negative interest rates, in other words, a type of income transfer between those that contracted the loans and the creditors. Leite (2001), regarding the origin of public resources for financing, affirms that they "came from the open accounts in the Monetary Budget and [...] counted on external fund raising and on the expansionist provisions from the Bank of Brazil".

Between the 1970s and 1980s, there was a large volume of investments benefiting from the state's financial intermediation that raised funds in the surplus economic units, generating a positive difference between revenue and expenses on the country's commercial budget, a fact that allowed it to reinvest the final balance back into its own financial system.

In 1979, for example, with the goal of attracting international capital, typically speculative capital, to bolster the country's international reserves, the U.S. manufactured a hike in interest rates to amplify its own reserves and protect the national currency, with this came the draining of international resources. According to Tavares (1999), the North-American decision, started in 1979–1980, "multiplied Brazilian foreign debt by three and brought on a foreign debt crisis for the capitalism dependent countries and the world crisis of 1980/81 for the planet".

> The investment resources disappeared the same time in which revenue from the exportation of *commodities* diminished considerably. Since the beginning of the 80s, the Brazilian Miracle came to an end, a period of accelerated growth that the country experienced, linked to world economic growth and to the abundance of foreign investment resources.
>
> PRETTO, 2005, p. 28

The *Green Revolution's* production model imported by Brazil and characterized by its lack of having an agrarian reform, with the technological transfer, and the limited training of the labor force for the urban capitalist cycle, categorically expresses the dependent development of Brazilian capitalism and the conservatism of the political pact of the national elite. Marini (2005), analyzing Brazilian development, affirms that international capitalism created a structural cycle of dependency and subordination, whose cornerstones for the relations of production were and are still modified or recreated to ensure the expanded reproduction of dependency.

Considering this historical process, the agrarian question started to be redeveloped, not only by conservative sectors, but also by a portion of the progressive sectors that spread the idea that there are no longer any latifundia and that waving the *flag* for agrarian reform is backward and unnecessary. In this regard, the reconfiguration of capitalism in the countryside, occurs with the adoption of neoliberalism after the revival of democracy and following agribusiness's precepts.

4   The Fight to Establish MST

MST was officially founded in 1984 during the fight for the return of democracy after 20 years of exclusion fueled by the military dictatorship and the dependent capitalist economy, that among other things, maintained subordinated agrarian development to the central economies and to the urban poles of development that drained the wealth produced in the countryside, impoverishing it (Fernandes, 1977).

Professor José Flávio Bertero (1999, p. 193) refers to the subordination of agriculture to the industrial urban sector, as Florestan Fernandes pointed out:

> Even when it, agriculture, achieved specializations that were consistent with the evolution of capitalism, obtaining consistent trends of technological modernization, it saw itself constricted within a national market that redefined "inside" as the same type of connection that the "outside" has. It is about what can be called "dependency within dependency" or "internal colonization". The expansion of capitalism, on the cornerstones of national economic development, was insufficient to inscribe a greater sense of autonomy on its agrarian economy.

Bernardo Fernandes and João Pedro Stédile (1999) points out that MST's genesis is linked, among other things, to the "dependency within the dependency" and to the particularities of the socioeconomic transformations that Brazilian agriculture suffered in the 1970s, rooted in an intense and fast process of capitalization and mechanization of Brazilian land tillage.

Since the beginning of the 1980s, the process of *modernization* of agriculture resulted in a large contingent of rural populations evicted from the countryside and serving as a cheap labor force for the urban-industrial and urban-commercial sectors. The economic system turned out to be fragile by the end of the "economic miracle" and the crisis devastated the industry and the Brazilian countryside. In this context, the process of colonizing the agricultural borders was a failure, since the *peasants* were not able to reproduce the production model of *modernization* nor that of the *Green Revolution* on their lands.

According to Stédile and Fernandes (1999, p. 17), this was the social context in which MST originated, because only a portion of the peasant families were willing to stay in the countryside, since they couldn't conceive that going to the city would be the solution to their problems, only leaving them the option to fight for the land and try to resist in the countryside.

The workers[4] that lost work and access to land as a process of *modernization* of agriculture, began to organize and fight to continue as residents and agricultural workers in the states in which they were born. Among these fights, we highlight the occurrences on the plots of land of Macali and Brilhante and the settlement of more than 600 families in Encruzilhada Natalino in the state of Rio Grande do Sul, where the fight for land started to gain new dimensions.[5]

Regarding this, Morissawa (2001, p. 123) highlights

> that the seed that would be MST was planted on September 7, 1979, during the military dictatorship, when the occupation of Macali Farm (Fazenda Macali) in Ronda Alta occurred, in Rio Grande do Sul. Many

---

4 Peasants, sharecroppers, tenants, squatters, farmers, ranchers, among others.
5 At this time, land occupations and encampments of landless families began to multiply in various Brazilian states, as was the case with the Burro Branco estate in the Campo Erê region of Santa Catarina, the occupation of the Annoni in Marmeleiro farm and the Mineira farm in São Miguel do Iguaçu in the state of Paraná. In Mato Grosso do Sul, the unemployed from the cities, who in this case were people who had been expelled from the countryside as victims of the military government's agrarian policy, also began to organize for the occupation. In the state of São Paulo, the struggle also began to emerge, as in the case of Primavera farm in the city of Andradina, from which the experience led to the formation of MST in the western region of the state, the struggle at Pirituba farm, located partly in Itapeva and partly in Itaberá, and the emblematic struggles in the Pontal do Paranapanema region are some examples of the context of the struggle for land in the late 1970s and early 1980s. For a more detailed reading, see Morissawa (2001).

other fights, in this State and in the entire country, garnered leadership and incremented the consciousness for the need to amplify their struggles in search for a larger objective: agrarian reform.

João Pedro Stédile mentions that the work done by the Pastoral Commission of Land (CPT) was also an important element for MST's genesis, due to its work of building consciousness with the peasants, that created space for the process of political education, without which, MST could not have been born, or rather, in the best case scenario would have taken a longer time to emerge.

CPT represented an organization of bishops, priests and pastoral agents, who during the height of the dictatorship contested, by means of Liberation Theology, the model of agricultural production that was being implemented in the countryside. The actors which comprised CPT discussed, with the rural workers, the need for them to organize, however, they didn't perform the typical messianic role nor the idolatry of "waiting because you will have land in heaven", on the contrary, they took on a position that fomented the organization of peasants fighting and resolving their issues here on earth. In addition to this, CPT's ecumenical and unifying role enabled the non-fragmentation of the rural workers in many organizations, contributing to the construction of a large movement and one on a national scale[6] (Stédile; Ferandes, 1999).

In the midst of this conjuncture, on the days 20, 21, and 22 of January 1984, on the 1st National Assembly of the Rural Landless Workers Movement, held on the premises of the Diocesan Seminary of Paraná, MST arises as a movement of workers and rural workers under the auspices of three priority demands: the fight for land, the fight for agrarian reform, and the fight for general changes in society.[7]

With the goal of nationally organizing the fight of workers and workers on the countryside, MST structured itself around three main principles: a) to be a large popular movement, with free access to everyone interested in the fight for agrarian reform; b) to be a union component, in the corporate sense that holds the interests of those of the working class and rural workers; c) to

---

6   João Pedro Stédile points to the struggles for the democratization of the country in its broadest sense as an important element in the constitution of the MST. He says that if these mobilizations against the dictatorship had not also taken place in urban centers, there would have been no conditions for the constitution of MST, so that the emergence of MST cannot be credited as the exclusive result of the will of rural workers.

7   The first meeting was attended by workers from 12 states: Acre, Bahia, Espírito Santo, Goiás, Mato Grosso do Sul, Pará, Paraná, Rio Grande do Sul, Rondônia, Roraima, Santa Catarina and São Paulo, as well as representatives from ABRA, CUT, CIMI and CPT (Morissawa, 2001).

be political, not restricting itself to the corporate aspect, so that the fight for agrarian reform becomes the constituent element of class struggle.

The occupations profoundly marked MST's first political acts and were the topic of discussion in the first two National Congresses of the Landless, held, respectively, in the years 1985 and 1990, which demarcated: a) the intense occupations with the intent to leverage agrarian reform;[8] b) the insertion of the movement in the 1st National Agrarian Reform Plan, boycotted by UDR's action.

In this sense, MST's tactic to occupy territory was a concrete and unifying action from the workers' fight which was not an isolated rupture or request for favors, but it was founded on the following words of order: *occupation is the only solution so occupy, resist, and produce*. This tactic confronted the elite, that could tolerate a request for favors and begging from the poor people, "but would never accept that they organize to demand their rights" (Stédile; Fernandes, 1999, p. 113).

Roseli Salete Caldart (2004), MST educator, discusses the theme of the teaching Landless individuals and highlights that "occupation can be considered the essence of MST because it is with it that the organization of these people to participate in the fight for land begins" (Caldart, 2004, p. 168), while Stédile (1997) affirms that within the occupation is what could possibly be called "the organizing foundation of MST".

This strong fighting stance and land occupation taken on by a part of the organized workers for and within MST forged the takeover of many territories intended for agrarian reform. The conquered territories under MST's organizing influence are made up of settlements that fought and fight to maintain and progress as much as it is possible to do so independently and not subordinated to agribusiness and to large capital. Although, this doesn't mean that the conquered territories are immune to the onslaught of capital and possible reconcentration, if no alternative socio-productive action was taken (Christoffoli, 2012).

It is in this sense that the catchwords of the 2nd National Congress, held between the 8th and 10th of May 1990, were directed beyond just the strengthening for the fight for agrarian reform, adopting key words such as occupy,

---

8 Morissawa (2001) points out that in the state of Santa Catarina alone, 5,000 families occupied around 40 farms. João Pedro Stédile points out that the number of occupations is not precise, but it is believed that in the first 15 years of the movement there were more than 1,500 occupations. The author also points out situations such as the São Bento farm in Pontal do Paranapanema, where it took 23 occupations before the government released the land for settlement.

resist, and produce, in other words, to give incentive to the agricultural production in alternative models, as a means of resisting the onslaught of capital. In this context, the movement expressed that beyond the occupation and the takeover of territories there was also a need to organize and strengthen production in the areas of the settlements already taken. The stance taken was to "develop cooperation as a form of strategic action in view of the advance of capital in the reformed areas, but also as a test run for the future organization of agriculture in a socialist society" (Christoffoli, 2012, p. 171).

João Bernardo (2012) describes that MST started to conceive of the cooperatives as a strategy to facilitate the access to financial and technical resources for the rural workers, and to create favorable conditions for production and commercialization, since agrarian policy penalized small farmers more than it enabled their survival. From the co-operatives, MST developed a conception of socialization for all of the factors and stages of production, such as land, capital, and labor. Inspired by the Cuban experience, MST moved on to a new phase, advancing debate and action in the conception of a cooperative system for Brazil, a national system with the role of meeting the demands of the different realities of the settlers in the country.

The organization of the cooperatives associated with the consolidation with agro-industries was aimed at putting agrarian reform products onto the market, believing that through these actions a qualification of production would occur and, consequently, a social and economic evolution for the families in the settlements. It should be emphasized that in this moment MST was fighting for a classic type of agrarian reform. For Nilciney Toná (2011), the movement believed that the ruling class possessed interest in placing the peasants into capitalist production as a means of complementarily, incorporating their productivity into industry.

At the end of the 1980s, MST created the Settler's Cooperativist System (SCA) and by the beginning of the 1990s, the number of agriculture cooperatives in the movement's settlements had increased considerably. This process resulted in the formation of the National Confederation of the Agrarian Reform Cooperatives of Brazil (CONCRAB), which in 1992 factored in approximately 55 production and commercialization cooperatives and 7 central state cooperatives. Along with this, more than 40 Agricultural Production Cooperations (CPAS) were organized, "many of them entirely collectivists, a true socialist enclave, not only in terms of work, but also in terms of certain aspects of domestic life such as, the use of cafeterias and child day-cares" (Bernardo, 2012).

The production cooperatives founded by MST had legal personhood so that they could insert themselves into the commercial circuit. Its organization consisted of planning, production, and of the creation of direct routes for commercialization, eliminating the tradition middlemen, in other words, it sought the autonomy of the settlements through the means of control over the chain of production that they developed (Bernardo, 2012).

Despite the euphoria, there were big challenges and contradictions imposed by capitalist logic:

> the unfamiliarity and the peasants mistrust facing these types of collectives resulted in a partial reversion of experiences, initially breaking down into semicollective groups and finally into the complete deconstruction of various complex cooperation initiatives. The discrepancy between the proposal conceived by the movement, of totally self-managed collectives, a lack of state support, insufficient technical training, and the derived contradictions of the organic organizational consciousness of the peasants were fatal for many of these experiments and forced a tactical retreat from the movement.
> CHRISTOFFOLI, 2912, p. 175

One can not ignore that the lack of comprehension and institutional judicial apparatus to enable self-management and collectivization for the means of production and labor in the cooperatives operated as a strong inhibiting agent and it was contrary to MST's proposal, which was ever more hampered by the obstacles that the state's action imposed on the cooperatives' training.[9]

In light of this, CONCRAB, since 1994, began focusing its efforts on providing services, stimulating the establishment of regional cooperatives and not only self-governed collectives. This model allowed more flexibility for the organization and unification of the families in the settlements that individually produced on their plots of land.

In this respect, the concept of cooperation in MST sought to transcend the complex question of production or even the legal bureaucratic organization, in order to maximize the political-ideological education necessary so that the settlers could participate in the fights and the demands of solidarity for other

---

9 The Collor government, for example, restricted credit and technical assistance for small-scale farming. It dismantled the Ministry of Agrarian Reform and Development, cut down the National Institute for Colonization and Agrarian Reform (INCRA) and used the Federal Police to repress MST, raiding state secretariats, seizing documents and arresting and prosecuting leaders. This period of repression led to a drop of almost half in the number of occupations, from 80 in 1989 to 49 in 1990, and a drop of almost half in the number of families mobilized, from 16,030 to 8,234 in the same period. In this difficult situation, having to survive on the defensive and relying mainly on its own resources, MST focused on developing production cooperatives (Bernardo, 2012).

categories that aren't only related to the worker in the countryside. This concept was part of the frame of reference for the 3rd National Congress of the Landless, held between the 24 and 27 of July, 1995, in which the key phrase was 'Agrarian reform is everyone's fight', seeking to express that the guidelines and actions of the agrarian reform could be amplified to levels beyond those of just the workers in the countryside.

It should be noted that in the beginning of 1995, Fernando Henrique Cardoso took over as president of Brazil and advanced the neoliberal agenda in the government. In this dynamic, despite the struggles carried out by the workers on behalf of the agrarian reform, the government demonstrated little concern for the concentration of landownership, while its main efforts in this area were to avoid the conflicts in the encampments from turning into a political problem (Morissawa, 2001).

Fernando Henrique Cardoso (FHC) strived to conduct the economic aspects of the agrarian reform by politically isolating MST and breaking apart its social base, for this purpose, Banco da Terra (Land Bank) was established in 1998, with the goal of replacing occupation with access to land through the mechanisms of the market. João Bernardo (2012) highlights that the strategy of supporting family farming, with the direct placement of families into the market by means of acquisition programs for farming production, occurred in detriment to the collective relationship held by the settlers' cooperatives, representing a strong positive impact for its clash with MST.

This action entailed a barrier for the Special Credit Program for Agrarian Reform (PROCERA), that ended up being dissolved in 1999, and was replaced in 1995 by the National Program for the Strengthening of Family Farming (PRONAF) which "ceased to be only a line of credit and was converted into a governmental program. It was a question of dismantling the production cooperatives, redirecting the credit to family farming" (Bernardo, 2012).

The replacement of PROCERA with PRONAF put the CPAs in big problems regarding financing, and in a certain manner, forced MST to stop favoring the more complex types of cooperative training and started submitting proposals from cooperatives linked to conventional forms of commercialization, breaking away from collective labor. Since then the direction of MST has given priority to the cooperatives that provide services.

By starting a new line to promote family agriculture with credit from PRONAF, FHC garnered a prominent strategic win in regards to MST, in a presidency in which the rest of it did not have any large successes (Bernardo, 2012). In this period, the sector of the agribusiness began to advance more distinctly in the national political-economic landscape. Since 1996, agricultural policy has been progressively conducted by an opening of markets geared towards import and to stimulating the input of foreign capital.[10]

Following this policy, the government spent more than 3 billion dollars importing food that could have been produced in the country. This policy resulted in a crisis that affected just as many small farmers as large-scale farmers. The production of cotton, for example, in which the country was the largest importer, fell to being third place, resulting in a loss of more than 400 thousand employed workers in the production chain (Morissawa, 2001).

MST, in turn, organized large mobilizations, among which we can emphasize the National March for Agrarian, Labor, and Social Justice Reform held in 1997.[11] "The goal of MST's national march was, in addition to calling attention to the urgency for agrarian reform and seeking punishment for those responsible for the massacres of rural workers, to celebrate International Day of Peasants Struggle for the first time" (Morissawa, 2001, p. 159). An important action and achievement for the struggle of the Landless was the National Meeting of Educators for Agrarian Reform, held in July of 1997, in the University of Brasilia (UnB). The meeting was organized by the education segment of MST in collaboration with UnB, the United Nations Educational, Scientific and Cultural Organization (UNESCO), and the United Nations International Children's Emergency Fund (UNICEF), with the aim of discussing education and agrarian reform.

Carrying out occupations, marches, and being inserted into other discussions that concern the improvement in the quality of life for the agrarian

---

10 Rural credit, which in other governments reached 15 billion reais, was just over 4 billion. Of the almost 4 million small farmers, only 168,000 obtained credit worth 200 million reais (1,190 reais per family). As for the interest rates, although still low compared to commercial rates, they represented a transfer of income to the banks by the small farmers, who were left without any income. With the overvaluation of the real against the dollar, aimed at making imports cheaper, and the elimination of customs tariffs, the market was flooded with imported agricultural products. The low prices of these products hurt domestic farmers and made it impossible for exporters to earn an income. The prices of products such as milk, pork, corn, poultry, etc., were pushed down in order to maintain the value of the basic food basket and, consequently, the minimum wage (Morissawa, 2001, p. 157).

11 Formed by three pillars departing from different places, one from São Paulo, another from Minas Gerais and the third from Mato Grosso, the march arrived in Brasília on April 17, exactly one year after the Eldorado dos Carajás massacre in which 19 people were killed by the police, as well as 69 wounded and 7 missing. The arrival in Brasilia was celebrated with a large public event attended by more than 100,000 people (Morissawa, 2001). João Pedro Stédile points out that the march also sought to confront the tactic of isolating MST proposed by FHC's government, in order to dialogue with society and confront FHC's government (Stédile, Fernandes, 1999).

reform population and for the society as a whole, MST continues to insist and take things head on in face of the government's inaction in implementing the agrarian reform. In this context, MST held the IV National Congress in the year 2000, in a moment in which there was a split with the technological package of the *Green Revolution* and the denial of agribusiness, which was gaining strength, in the face of the consolidation of neoliberal policies in Brazil. The catch phrase of the IV MST Congress was: "For a Brazil without latifundia", in which the movement begins to envision agricultural production as an alternative to agribusiness.

Since then, MST began to make harsh critiques against the *Green Revolution* and agribusiness, among which we can highlight: the occupations of transgenic seed and pesticide factories; mobilization against the patent policies; complaints of the environmental and human evils brought about with the use and consumption of feed contaminated by pesticides, etc.

In this view, the fight for agrarian reform becomes more complex, given that the enemy isn't only large unproductive estates anymore, but agribusiness and the entire structure that maximizes on the exploration of workers and natural resources. On this level, agroecology is assumed as the fundamental strategy to overcome the predatory model of agribusiness.

Agribusiness, in addition to operating under the abuse and control of foreign investments in production and the agricultural markets, began to have a significant economic role within financial capital's activities, generating commercial balances to increase the foreign exchange reserves, an essential condition to attract speculative capital to Brazil. This advance of agribusiness, according to MST's analysis (2012, p. 12) "blocks and protects unproductive land for the future expansion of its businesses, hindering the acquisition of land for agrarian reform".

CHAPTER 5

Perspectives and Dimensions of Agroecology

1 Introduction

Agroecology, or more precisely, the constitution of the agroecological framework was constructed under the influence of the agriculture sciences, with a conceptual, methodological, and practical interaction with ecology, sociology, and anthropology, and geography which provided a rich intellectual contribution to "the social impacts of technology, the pernicious effects of the market of commodities, the implications on the changes within social relations, the transformations in the structure of land possession and the growing difficulty of accessing common resources for the local population" (Hecht, 2002 *apud* Moreira, 2003, p. 11).

In our studies, two names stand out as possible proponents of the agroecology concept, one is a Russian agronomist Basil Bernsin (1928) and the other, English Albert Howard (1930). For both, agroecology was proposed with a foundation in ecology studies applied to agriculture. However, the two areas of knowledge (ecology and agronomy) which, at first, gave the theoretical and practical base for agroecology, didn't have a fully integrated relationship during the following decades.

Gliessman (2002) emphasizes that this poor integration took place, among other things, by the means of development from the sciences connected to the land in the capitalist mode of production. Modern soil science, developed with a base in the introduction of chemical fertilizers and addition of nutrients to the soil, ignored the need to develop ecological practices for regenerating soil fertility. Foster (2010) highlights that this measure, since the beginning, presented some dramatic results for the productivity of the soil, something proven by the fact that soil fertility in general is always limited to the less abundant nutrient (the law of the minimum, by Liebig).[1]

In this dynamic, the ecological question was restricted, almost exclusively, to the studies on natural systems, those without anthropogenic intervention. On the other hand, agronomy, equipped by modern soil science and influenced by

---

1 Justus Von Liebig was a German chemist who, in 1840, formulated the Law of the Minimum while studying plant growth. The Law of the Minimum = Liebig's Law described that under steady-state conditions, the nutrient present in the smallest quantity (concentration close to the minimum required) tends to have a limiting effect on plant growth.

the capitalist production logic was left with a hegemony of studies connected to agriculture production. Developed under the influence of the industrial-capitalist mode of production, modern soil science was guided by an ethos of reductionism and utilitarianism that aimed to increase the level and velocity of productive soil fertility, ignoring its natural need for recomposition.

Foster (2010) indicates that Marx, rooted in the historical events of the time, had already pointed out a possible degradation of the natural resources generated by the excessive use of the soil, such as:

> (1) the growing sensation both European and North-American of the the agricultural crisis associated with the depletion of natural soil fertility – a sensation of crisis that absolutely wasn't alleviated, but in fact driven, by the advances of soil science; and (2) a turning point in Liebig's own work at the end of the 1850s and in the 1860s in a direction towards a strong ecological critique of capitalist development.
>
> FOSTER, 2010, p. 213

Foster (2010) describes that the degradation of the soil, the destruction of the natural nutrient cycle, the growing knowledge of the need for specific nutrients, a fragmented vision of agricultural production, the limits of supply for natural and synthetic fertilizers contributed more and more to a generalized sensation of crisis.[2]

The considerable scientific and technological development showed itself to be incapable of recuperating and maintaining the necessary conditions for the recycling of the constituent and essential elements of the soil, given that capitalist logic of constant productivity required its intensive usage, leading to excessive wear and tear, without enough time for it to recuperate (Foster, 2010). Considering the limits of modern soil science and capitalist rationality in agriculture, Gliessman (2002) describes that by the end of the 1920s the field of crop ecology merges. In the following decade, searchers connected to that

---

2   The contradictions of agriculture in this period were felt particularly acutely in the United States, which suffered from the British monopoly on shipments of Peruvian guano (rich in nitrogen and phosphate), which blocked easy and economically viable access to the product. This led the United States, through a state policy, to undertake the imperial annexation of any island that was believed to be rich in this fertilizer, however, the imperialism of guano did not enable the United States to obtain the necessary quantity and quality of these fertilizers (Foster, 2010).

branch of ecology begin to utilize the concept of agroecology to specify the application of ecology to agriculture.[3]

Despite the fruitful discussion on the need to develop a less degrading agricultural production for the environment, the advancement of the capitalist production model in agriculture and the fragmentation of knowledge confined ecology to the study of natural systems, leaving agroecology, as a branch of less importance in the field of agricultural sciences.

It is still necessary to consider that the discussion regarding agroecology occurred between a period of time where there was two wars, enhancing technological development. After the Second World War, large multinational corporations, with headquarters in the countries with large production of capital assets (United States, Great Britain, and Japan, for example), that played a role in the scientific and technological development of the industry of war, adapted a part of their area of expertise to enhance the process of capitalization of agriculture and agricultural modernization[4] (Harvey, 1993).

In the first decades of the second hand of the twentieth century, the ecological question continued to not be a priority, with the modernization of agriculture based on the reductionist and utilitarian scientific advances prevailing, that stimulated the usage of chemical fertilizers, pesticides, introduction of machinery, production on a large scale, and monocultures, as a means to create large-scale production, increasing farm profits. Large-scale agricultural production also consumed more energy, making the addition of a large amount of energy necessary, which required not only the use of electricity but also use of its dissipation in the form of heat, which originated from non-renewable sources of energy, then being converting into contamination in the soil, air, water, in other words, the accumulated waste in the environment (Moreira, 2003).

Since the 1960s, due to the consequences of the contradictions that the conventional model of development and utilization of natural resources had already been generating since the second half of the 19th century, the discussion

---

3 Sir Albert Howard is considered to be the founder of the organic revolution in agriculture, as well as one of the founders of the concept of agroecology. His main research was carried out in India, where he studied peasant farming culture in depth. His work "An Agricultural Testament", first published in 1943 in London by Oxford University Press, was made available in Portuguese by Editora Expressão Popular [Expressão Popular Publisher] in 2007.

4 A fact that illustrates this is the use of Agent Orange, also known as 2-4-D, which was used by the American army in the Vietnam War to spot the enemy on the ground, as its application causes plants to defoliate. After the war, it was adapted as a herbicide for use in agriculture. It should be noted that exposure to it not only causes various diseases such as cancer, but also contaminates fauna and flora.

of the so-called *environmental consciousness*[5] begins to gain momentum in the main capitalist countries. In this context, Gliessman (2002) describes that the progress of the studies on the environment made the consolidation of the concept of ecosystem possible, providing, for the first time, a considerably coherent reference point to examine agriculture from an ecological perspective.

> an ecosystem can be defined as a functional system of complementary relations between the living organisms and their environment, delimited to arbitrary criteria, of which in space and time seem to maintain a dynamic balance. Thereby, an ecosystem has physical parts with special relationships, the *structure* of the system – which, in its conjecture, form part of the dynamic processes – the *function* of the ecosystem.
> GLIESSMAN, 2002, p. 17

The ecosystem is the main area of ecology studies. It can be represented as a natural area composed of living beings (biotic environment) and the location where they live (abiotic environment where the non-living components of the ecosystem are inserted, such as atmospheric gases, mineral salts, and solar radiation). Studies on ecosystems boosted the interests in crop ecology once again which reappear with the nomenclature *agriculture ecology*. In the year 1917, in Holland, the 1st International Congress of Ecology was held, where a group of researchers presented work on the analysis of agroecosystems. An agroecosystem is an artificialized ecosystem, a site of agricultural production that serves as a reference to analyze systems of food production in their totality (Gliessman, 2002).

The concept of agroecosystems established a strong basis for analyzing agroecology. Casado; Sevilla Guzmán and Molina (2000) interpret the agroecosystem as a artificialized ecosystem by human practices by means of the systems of knowledge, social organization, cultural values, and technology. Its internal structure results from the social relation, the product of the *coevolution* between human societies and nature.

The concept of agroecosystems was constituted in an analysis perspective from the agricultural activities performed in small geographical areas. They are

---

5   From the 1960s onwards, concern about the degradation of the conventional development model intensified. Works such as Silent Spring (1964), by Rachel Carson, strongly questioned the secondary effects caused to the environment by the use of toxins in the Green Revolution model. The work of G. Douglas (1984), "Agricultural sustainability in a changing world order", was also important in solidifying the relationship between agroecology and sustainable agriculture.

understood as open systems that receive external inputs that generate products that can be exported beyond their borders. However, one must consider the complexity of precisely defining what an ecosystem is (Altieri, 2012). The studies that were emerging on agroecosystems and the socioenvironmental imbalance, since the 1970s, generated a basis for which agroecology reemerged as an alternative in the scenario of studies about the scientific and technological framework of agriculture.

Moreira (2003), in his study on the processes of transition from the conventional model to an agroecological system shows us two schools of thought: a) the North-American branch, that has its origins with researchers in the state of California and b) the European school, with emphasis on the Institute of Sociology and Peasant Studies (ISEC) from the University of Cordoba in Spain. Utilizing Moreira's work (2003) as a reference, it turns out that the North-American school follows a theoretical approach, connected to the technical issues and sustainable management, as an alternative to the degrading practices and pollutants; while the European branch goes beyond the technical and management questions (without abandoning them) of a social theory and critique of capital's logic.

The presence of both of these branches is evident when analyzing the theoretical references used in MST's agroecology courses, therefore, these two currents are presented in the next two subheadings, in order to explain which of them resounds the most with the movement's discourse.

2   The Perspective of the North-American Thought: For Sustainable Processes in Agriculture

Professor Stephen R. Gliessman is one of the main exponents on agroecology from the North American school. He emphasizes that agroecology's objective is to create agricultural management practices founded on environmental equilibrium, sustainable yields, soil fertility through ecological processes and natural regulation of pests, design of diversified agroecosystems, and the use of low external input technology (Gliessman, 2002). From the studies, connected specifically to traditional Mexican agriculture, Gliessman (2002) makes a powerful critique of the utilitarian and reductionist development model of agriculture, that increases the consumption of energy coming from (directly or indirectly) non-renewable sources (fossil fuels), in order to produce more.

In addition to the dependence on external non-renewable energy, the conventional model of agriculture causes damage that is expressed in: a) decrease in soil fertility; b) loss of organic material; c) leaching of nutrients;

d) degradation and increase in soil erosion e) contamination and depletion of water sources; f) increase in pests and diseases; g) contamination of the agricultural environment and natural ecosystems; h) damage to the health of the farmers and agricultural wage earners; i) destruction of beneficial insects and microorganisms; j) drastic reduction of the regional biodiversity; k) imbalance in the global nitrogen cycle, which leads to the aggravation of the problems in the ozone layer (Gliessman, 2002).

Seeking to establish the guidelines of sustainable agriculture, built on the basis of the conservation of human and natural resources, agroecology can be defined as the "application of concepts and ecological principals to design and management of sustainable agroecosystems" (Gliessman, 2002, p. 13).

> the agroecological focus begins paying attention to a specific component of an agroecosystem and its possible management alternative, during the process it establishes the basis for many other things. Applying the focus, in a broader manner, allows us to examine the historical development of the agricultural activities in a region and determine the ecological basis for selecting more sustainable practices for the ozone. Also, it it can help us find the causes of the problems that have emerged as a result of unsustainable practices. Nevertheless, the agroecological focus helps us to explore the theoretical bases for the development model that can facilitate the design, tests, and the assessment of sustainable agroecosystems. Finally, the ecological knowledge of agroecosystems' sustainability, should restructure the current focus of agriculture with the goal that humanity provide sustainable systems of food production.
>
> GLIESSMAN, 2002, p. 13

Agroecology seeks, through the ecological management of agroecosystems, to develop sustainable systems of food production. Miguel Altieri (2012), a well-known researcher linked to the University of California, asserts that the concept of sustainability is polysemic, although a consensus has been generated in relation to the need to propose modifications of conventional agriculture, in a way that makes them more viable and compatible from the environmental and social point of view.

Establishing the basis for the development of sustainable alternatives for agroecological agriculture, Gliessman (2002, p. 217) defends polyculture in replacing monoculture, explaining that the population of mixed plants are capable of *coexisting* given the variety of mutual adaptation mechanisms, such as the distribution of physiological behavior that reduces competition among the crops. Agroecology, in the words of Altieri (2012), should extrapolate the unidimensional vision for the agroecosystems (genetic, edaphology, among others), since it sees the complete integration among the ecological and social levels of *coevolution*, emphasizing the inter-relationships of its components and the complex dynamic of the ecological processes.

With a foundation in Gliessman (2002), it is considered that the concept of agroecology is based on a holistic aspect, that opposes the *Green Revolution's* proposal, by structuring itself on a foundation in the following guidelines: a) low dependency on commercial inputs; b) use of locally accessible renewable resources; c) utilizing the beneficial or benign impacts from the local environment; d) accepting and/or tolerating the local conditions before dependence on the intense alteration or attempt to control the environment; e) long-term maintenance of the production capacity; f) preservation of biological diversity; g) utilizing the knowledge and culture of the local population; h) production for internal consumption before the production for exportation.

The North American agroecological perspective is founded on the sustainable management of agroecosystems, with the application of the following practices: a) usage of solar and wind energy; b) biological control of nitrogen and other nutrients, products for the decomposition of organic material or for the soil mineral reserve; c) crop rotation dynamics; d) green fertilization; e) organic waste, among other renewable resources (Altieri, Nicholls, 2003).

Opposing the *Green Revolution's* agricultural production model on agroecology, it could be observed that there isn't a pre-made formula and it is being developed on the basis of theoretical and practical work, seeking to give value to the agricultural workers' empirical knowledge in the sense of an equitable relationship among all of the parties involved (Gliessman, 2002).

Agroecology, in the North American branch, seeks to rescue traditional systems that have proven to be sustainable in ecological terms. Agricultural systems are designed as a *coevolution* that occurs between culture and environment, where human beings are capable of mediating this *coevolution*. In this sense, to revert the predatory process of the conventional model of agriculture, agroecology should be developed from a perspective of totality, considering the interdependent relations between human beings and the environments in which they live.

The North American school has been creating alternatives for the development of sustainable agriculture and, with this, it has also contributed to the development of the epistemological, scientific, and technical concept regarding agroecology, establishing a new basis for food production that is less degrading to the environment and to the human being. In addition to the critique of the *Green Revolution* and to adopting agricultural management practices to preserve the environment, the North American model presents limits, mainly because it focuses more on productive-technical pragmatism and on the sustainable management of natural resources than having a critique of the structure of the model of capitalist agricultural production. Thus, it leaves some gaps in relation to the political, economic, and cultural struggles, for example, that make up the contradictory scenario of agricultural development in a capitalist state.

This in light, the North American agroecological branch only seeks to readapt the predatory model of conventional agriculture to a more sustainable framework, confining itself to a reformist logic, that fails to radically question the production rationale of capital.

3   Resistance and Existence: Agroecology in Spanish Thought

Eduardo Sevilla Guzmán and Manuel Gonzáles Molina are the main exponents on studies on agroecology in Spain, who founded the Institute of Sociology and Peasant Studies (ISEC) at the University of Córdoba in Spain. For this school, agroecology develops by means of social agrarian thought, of the movements opposing and resisting the industrialization of agriculture, characterizing itself as a constant dialectic between capitalist modernization and the resistance to its structural foundations (Sevilla Guzmán: Woodgate, 2013, p. 27).

Moreira (2003), analyzing agroecology in Spanish thought, iterates that it is constructed through the interactions between the scientific disciplines and the empirical knowledge of the rural communities and their traditions, in particular in Latin America. Something that perhaps allows us to understand why its expansion in Spain occurred in those regions where agrarian modernization had taken place late, as it is in the case of Andaluzia[6] (Moreira, 2003).

Agroecology in Spanish thought, like the North American concept, is also founded on a holistic aspect, in dialogue with different theoretical orientations that oppose conventional agricultural thought. Despite being rooted in the natural sciences, agroecology combines, in a more profound manner, with sociology, possessing a more critical vision in relation to political economy on

---

6 By the end of the 1980s, Andalusia had a reality in which situations typical of a recent and territorially incomplete agrarian modernization were combined with all the environmental problems characteristic of post-industrial societies. This coincidence favored the emergence of the first agroecological studies at the Universities of Córdoba and Granada, and more specifically at the Institute of Sociology and Peasant Studies (ISEC) at the University of Córdoba (Moreira, 2003, p. 11).

agricultural production, at the same time in which it also respects the cultural knowledge constructed by rural workers. (Sevilla Guzmán; Woodgate, 2013).

Sevilla Guzmán and Ottmann (2004), Sevilla Guzmán and Woodgate (2013); Casado and Molina (2006); Sevilla Guzmán and Molina (2013) describe that agroecology, in addition to the ecological management of the natural resources, reassembles the emerging proposals of its endogenous potential and the collective action that present resistance and alternatives to the current model of predatory management of natural resources.

> To develop such a task, agroecology introduces, together with scientific knowledge, other forms of knowledge. It develops, therefore, a critique of scientific thought in order to generate, from it, a pluriepistemological approach that accepts sociocultural biodiversity. The empirical evidence obtained during the last four decades since agroecology's inception (Altieri, 1985; Gliessman, 2002; Casado, Sevilla Guzmán and Molina, 2000) has demonstrated that the accumulated knowledge on the agroecosystems in the past can provide specific solutions for each place to resolve their social and environmental problems. Even more so if different ethnicities (with different cosmovisions) have interacted with it at each historical moment, and have contributed their knowledge in order to obtain these solutions. Multiple forms of knowledge exist within the historically subordinated groups and this knowledge can be recuperated and incorporated into the design for the agroecological strategies.
> SEVILLA GUZMÁN, s/d, p. 1

According to Sevilla Guzmán and Woodgate (2013, p. 27) agroecology

> promotes the ecological management of the biological systems through the collective forms of social action, that redirect the course of co-evolution between nature and society with the goal of dealing with the "crisis of modernity". The aim is to achieve this objective through systemic strategies ... to change the mode of human production and consumption that have produced this crisis. For these strategies, the local dimension is fundamental, in which we find endogenous potential encoded within the systems of knowledge ... that show and promote as much cultural diversity as ecological diversity. This diversity should form the starting point for agricultural alternatives and for the establishment of dynamic and sustainable rural societies.

The scientific and technological hegemonic paradigm, from the perspective of development and progressive of the rural areas, caused hegemony in the agroindustrial model, that, among other things, generated erosion of local traditional knowledge. This erosion took place with a foundation in the process of a gradual imposition from the economic, social and political agendas linked to positivism regarding the agrarian sciences and the life model that was developed with capitalism.

Sevilla Guzmán and Woodgate (2013) reinforce the importance of coevolution between society and nature in agricultural production, which values the strategies of socio-productive organizations built on the local dimension, by means of social-collective action, given that they can be utilized as a means of overcoming the socioenvironmental crisis, overcoming the paradigm of the Green Revolution. Moreira (2003) highlights that despite social and environmental systems historically having constantly *coevolved*, it doesn't mean that this relationship may had been constructed beneficially as much for the populations as it had been for the environment. The m*odernization* of agriculture, imposed by capital's logic, via the *Green Revolution*, entailed that the metabolic relationship between society and nature has followed a disastrous trend.

According to Casado; Sevilla Guzmán and Molina (2000, p. 92) quoting Geogescu-Roegen, the function of production, utilized for agricultural modernization, looks like "a list of ingredients that is comprised of certain products and doesn't take into account the cooking time", or, in other words, it follows a mechanistic logic of production that appears to leave out the dimension of "time", which refers to the fact that natural resources are finite.

The productive logic of the *Green Revolution,* when thinking about pests in an isolated way in the system (atomism), advocates for the application of pesticides, which, in turn, *coevolve* with the pests, interfering even more, in the ways in which the farmer sees the production process. Indeed, the *Green Revolution* influenced and accelerated the process of *coevolution*, introducing multiple technological changes that most of the time don't fit the social and environmental complexity of the rural communities (Noorgard; Sikor, 2002).

The *coevolutionist* perspective, in a holistic sense, like the Spanish branch, seeks to revert the metabolic issue of socioenvironmental wear and tear, a consequence of the hegemonic model of the *Green Revolution*. Instead of following the *Green Revolution*'s hierarchical hegemonic model, it proposes placing the communities and their form of organization at the center of the process. In this light, there is respect for the knowledge regarding agricultural production that traditional or native communities have historically constructed, through a process of trial and error, and selection, in which they learned to capture the potential of ecosystems, preserving them (Casado; Sevilla Guzmán; Molina, 2000).

One of the most important characteristics of the *coevolutionist* approach is that it gives legitimacy to the farmer's knowledge, because, contrary to that of the external agents (scientists, researchers, extensionalists); they *coevolued* with nature in a more harmonic manner and improved, in many cases, their systems of production throughout the millennia. The Spanish approach is based on respect of the farmer's wisdom and brings attention to the fact that their scientifically encoded knowledge isn't the only source of legitimate knowledge (Noorgard; Sikor, 2002).

Noorgard and Sikor (2002) explain that social and environmental change can be carried out with coevolution between natural systems (climate, land, biodiversity, etc) and the social systems (values, forms of knowledge, and technology). For this reason, coevolution between society and nature should be based on interdependence and not on environmental and cultural determinism.

However, Spanish thought considers that the structures (organization and design) of the agroecosystems, in agricultural production, should not be materialized in isolation from social activity. Because of this, agrosystems need to be planned from the methodological procedures that stimulate agricultural workers to be included in the building and participation of alternatives, oriented towards the necessities of each *locus*, from a perspective of exchanging knowledge. Sevilla Guzmán and Molina (2013) calls the connection between social agrarian thought and the resistance movements to *modernization* of agriculture "alternative social agrarian thought", which it is based

> on discourses that, consciously or unconsciously, are behind collective actors who we call here "theoretical orientations" as intellectual categories, in which explanations and values are articulated on some level of reality, generating a process of legitimacy or delegitimacy of certain parts of that reality, in this case, relative to the peasant, agriculture, or rural society.
> SEVILLA GUZMÁN; MOLINA, 2013, p. 15

The theory of "alternative social agrarian thought" follows the holistic perspective of history, seeking to encompass a complex diversity of manifestations in a methodological strategy that has roots in the processes generated from historical identities that were constructed by social memories, coming from the perspective of the defeated (Sevilla Guzmán; Molina, 2013). Its approach values and reclaims the knowledge of traditional and/or native peoples, peasants, and the indigenous who correspond to those who Michael Foucault calls the insurrection of "subjective knowledge".[7]

In this regard, its development crosses the lines of the historical content made from "multiple forms of cultural resistance (starting with open rebellion and protest movements, to day-to-day forms of passive resistance, and the different systems of political domination)" that have formed values that were incorporated into the social memories that agroecology must rescue (Sevilla Guzmán, 2011, p. 16). With a foundation in "alternative social agrarian thought", Sevilla Guzmán and Molina seek to deepen Marx's afterthought in respect to pre-capitalist social formations. They seek to also rescue the potential of the peasantry and the indigenous and/or traditional populations for a possible transition to socialism, from the fight and resistance to capitalism, considering the "advantages of outdateness", and considering agroecology as a possible driving force.[8]

Developed from "alternative social agrarian thought", agroecology aims to question the ethnocentrism of the agricultural and social sciences that centered their research on a single, exclusionary civilizing proposal of social formation. It is also a proposal to modify and rescue, among other things, "peasant and indigenous" practices, that have historically demonstrated alternative and sustainable forms of socioproductive organization (Sevilla Guzmán, 2011, p. 12).

> The agroecological focus appears as a response to the logic of neoliberalism and economic globalization, as well as to the guidelines of conventional science, whose epistemological crisis gives rise to a new epistemology with participative and political character. And this is in the sense of "reinterpreting the question of power, inserting it in the ecological model, from which it follows that the real sphere of power is the social – as a living organism, as an ecosystem. It is the confrontation between the model of the artificial, closed, static and mechanistic system (the state); and the dynamic and plural model of the ecosystem (society)" (Garrido Peña, 1993, p. 8). The sociopolitical dynamic of agroecology moves in relational forms with nature and society, which Joan Martínez Alier defines as "ecological power", as a defense to its ethnoagroecosystems through

---

7 Focault, Michel. Society Must be Defended: Course at the Collège de France (1975–1976). São Paulo: Martins Fontes, 2005.
8 For further reading on alternative agrarian social thought, see Sevilla Guzmán (2011); Sevilla Guzmán and Molina (2013) and Sevilla Guzmán and Woodgate (2013).

> the distinct forms of peasant conflict in the face of distinct types of aggression from "modernity".
>
> MARTÍNEZ ALIER, GUHA, 1997 *apud* SEVILLA GUZMÁN, 2011, p. 16

Based on the guidelines of "alternative agrarian thought", agroecology is a means of questioning the legal framework of agrarian policy influenced by the technological paradigm of the *Green Revolution* and agribusiness, breaking away from the traditional concept of development and *modernization*. Sevilla Guzmán (2011) highlights that it is possible to develop socioeconomic production processes in the field based on participative methodologies capable of transforming the structures of local power, from analyses that allow establishing alternative proposals to technological development and agricultural productivity that isn't based on the degradation of the environment nor the exploration of man by man.

In order to do this, it is necessary to make a connection among the social collective action of ecologists, feminists, pacifists, in short, from the social movements with a history of struggle linked to the working class, including peasants, indigenous, and the water and forest communities, that resist the hardships caused by *conservative modernization*. In this sense, the historical content theorized regarding social action and agroecology, developed within and outside scientific thought, need critical awareness and a connection to the working class and to social movements, that fight and resist the demands of capital, because they are indispensable for the development of class consciousness, identity, gender, and generations, without which the power of minority resistance is limited.

4   Agroecology in the Brazilian Scenario

The intensification process of agricultural *modernization* in Brazil took place under the aegis of the military dictatorship which began in 1964 and continued until the 1980s, when political attrition pointed to a democratic opening that was realized after long-standing pressure from many areas of society, with the constituent process that culminated in the Constitution of 1988.

Since the end of the 1970s, many segments of society had already been mobilizing against the military dictatorship and for a constitutional state. According to Nunes (1989), Benevides (1994) and Schiochet (2012), these mobilizations provided the organization for portions of civil society in the social movements, which contributed to broadening the struggle for participation. Schiochet (2012) emphasizes that the contradictions and conflicts in the urban areas instigated the mobilization of workers in many social movements, such as: the cost of living movement, the housing movement, the fight against unemployment movement, the collective transportation movement, and the health movement. In addition to these movements, a movement occurred around "Direct Election Now" (*Diretas Já*) and the establishment of the Central Workers' Union (CUT).

In the field of education Dal Ri and Vieitez (2013) highlight that the fight for a constitutional state was led and organized mainly by the National Forum in Defense of Public Schools (FNDEP) that proposed democratic management as the fundamental principal for the organization of public schools, which occurred in formal terms with the enactment of the 1988 Constitution and the Law of Guidelines and Bases of National Education n° 9.394, from December 20, 1996.

On the agrarian question, MST was an emblematic example of an organization of thousands of workers that utilized marches, occupations, and settlements as methods to denounce their exclusion from the access to land and even precariousness of life in the countryside.[9] Despite the social demonstrations for political participation in the state and for the approval of the Citizens Constitution of 1988 which has a social welfare element, the democratic process initiated in the 1990s took place under the ethos of neoliberal policies, a reduction of the state and privatizations of many companies and public services that specifically follow the approved guidelines in the *Washington Consensus*.[10]

During the redemocratization (a process of restoring democracy) process in Latin America and in Brazil, in particular, the *environmental question* emerged alongside the agricultural workers' historical fight for the right to land and for the right to decent work and living conditions in the countryside. Simultaneously, with the social movements' fighting in defense of the

---

[9] As well as fighting for social and political rights, the social movements questioned the capacity of conventional trade unionism, which in practical terms was considered to be timid and ineffective in representing the different forms of rural workers. In addition to MST, the Regional Commission of People Affected by Dams, the Movement of People Affected by Dams, the National Council of Rubber Tappers, among others, also emerged. For further reading, see Medeiros, 2014.

[10] The Washington Consensus is a set of measures formulated in November 1989 by representatives of the IMF, the World Bank, among other financial institutions located in Washington D.C., which, in the case of Brazil, should be thought of within the limits of the restructuring of capitalism, the abandonment of nationalist pretensions and the readjustment of the Brazilian state to the new demands of accumulation and world capitalism.

environmental question, an international conference Eco 92 and/or Rio 92 was held, in the city of Rio de Janeiro where the establishment of Agenda 21 was one of its main outcomes.[11]

This concern with sustainability and preservation of the environment, even at the hand of reformist or neoliberal adaptation, gave new precedents to questions directed towards the *Green Revolution* model and its damaging consequences to socioenvironmental sustainability. The *environmental question* had already been seen coming by some segments of civil society and national and international Non-Governmental Organizations (NGOs)[12] in the course of the redemocratization process of the country in the 1980s. In this period, agroecology was presented by the NGOs as an alternative to family agriculture, given the difficulties faced by small farmers as a result of the way in which modernization of agriculture took place through the *Green Revolution* perspective.

In this scenario, researchers, extensionists and intellectuals, such as Ana Primavesi, José Antônio Lutzenbergerr, Ignacy Sachs, and Miguel Altieri stand out as the authors of papers on ecological management of natural resources and broader critical analysis. They demonstrated the limits of the degrading development of the *Green Revolution* and the need to reassess and modify the

---

11   The United Nations Conference on Environment and Development (UNCED), held in June 1992 in Rio de Janeiro, represents a milestone in the way humanity relates to the planet. Known as Rio-92, Eco-92 or Earth Summit, it took place 20 years after the Stockholm conference in Sweden. It was at the Rio 92 Conference that countries recognized the concept of sustainable development and began to establish actions aimed at protecting the environment. The so-called Agenda 21 was an instrument for adapting to the new form of globalization of capital and was defined as a planning instrument for building sustainable societies through methods of "environmental protection, social justice and economic efficiency". For more information on Agenda 21 and its developments, visit the website: http://www.mma.gov.br/responsabilidade-socioambiental/agenda-21.

12   For a more complete reading on the role of NGOs in the development of agroecology, we recommend following the work that Thelmely Torres Rego has been developing under the guidance of Professor Célia Regina Vendramini in her doctoral thesis at the Federal University of Santa Catarina. The qualifying text, which was presented in April 2015, analyzes the Alternative Technology Project linked to the Federation of Organs for Social and Educational Assistance – FASE (PTA-FASE), which aimed to disseminate alternative technologies in order to make family farming viable. This project resulted in the creation of the Alternative Technologies Project Network (Rede PTA), which brought together various NGOs with the same focus. One of these NGOs is Family Farming and Agroecology (AS-PTA), created in 1989 and founded as a non-profit civil law association in March 1990. With the growth of organizations, networks, projects and programs focused on the same theme, in 2002 the National Articulation for Agroecology (ANA) was formed, which, like PTA-FASE and the PTA Network, plays a fundamental role in the dissemination of agroecology in Brazil.

production framework, mainly of the dependent countries. This moment gave rise to scientific and technical contributions concerning the necessity and possibility of an ecological management of natural resources as an alternative for agriculture, with the emergence of agroecology emerging in North America and Spain, as we explained previously.

In Brazil, agroecology underwent a co-opting process from the international organizations such as the International Monetary Fund (IMF) and the World Bank, making eco-*developmentalist* strategies disconnect from social struggles, constituting a form of development that involves the "capitalist appropriation of the environment, its introduction into the production process for more value, and its introduction as a good in the market logic" (Sevilla Guzmán, 2011, p. 117).

In this sense, what occurs in Brazil is the development of a sort of *green capitalism* with prospects for a select market of organic products and/or with socioenvironmental responsibility, which is close to that which we present as the North-American branch. Moreover, the agroecological branch isn't the only branch that is propagated throughout Brazil. The advance of the environmental topic and ecological management of natural resources in agriculture has it that the Spanish branch has also gained notoriety in the development of a critical analysis of historical specificities for the *modernization* of Brazilian agriculture.

We highlight Pinheiro Machado, Francisco Caporal, José Antônio Costabeber and Enio Guterres as the authors that take on a more critical analysis on the historical particularities of Brazilian capitalism and the contradictions of the *Green Revolution*, proposing an agroecology that is more connected to struggle and social resistance while establishing a dialogue between the social movements from the countryside and small farmers. Guterres (2006) argues that the progress and propagation of the agroecological framework from the critical perspective, with transdisciplinary, theoretical, and methodological emphasis, require a more active stance from the academy, in the sense of exceeding the limitations of the North American branch.

Machado and Machado Filho (2014) point out that agroecology should be developed as a praxis of struggle and resistance on the part of the small farmers which should establish another paradigm for the management of natural resources and clean food production, as a means of overcoming the monoculture model of *modern* agriculture, inspired by the *Green Revolution* and agribusiness. This second group of researchers develop their work and research from a more critical vantage point of the totality of capitalism, not restricting themselves to a reformist stance that compiles with the proposal of *green markets* and organics. We consider that some of these authors are closer to the studies developed by ISEC.[13]

Caporal and Costabeber (2004) highlight that agroecology must be connected to the resistance struggles of rural people, constituting a scientific and technological framework that lays down the foundation for sustainable agricultural development, while respecting the knowledge and experiences accumulated by the rural workers, in their spaces of resistance. In this light, agroecology is much more than a theoretical debate although it is necessary; for this reason, we highlight the different meanings and descriptions regarding the subject. However, we stress the importance of not falling into a neverending epistemological discussion that may go nowhere, and, worse, even if the intention is to counter agribusiness, in practice, it ends up benefiting it, like the coopting that has been occurring with *eco-developmentalism* through the *green market*.

5   Agroecology from the Perspective of MST and Social Struggles

Agroecology began to gain momentum in the Latin-American scenario, and, consequently, in Brazil by the 1980s. Since then, many researchers, extensionists, and organizations have been theorizing its practices and principles, to the extent that there isn't a singular nor homogeneous definition of agroecology. It has been adopted as an alternative to tackling exclusionary conditions that agricultural capitalization created for many workers who are produced and reproduced in the countryside.

Based on Gliessman (2002), Moreira (2003), Caporal and Costabeber (2004), Sevilla Guzmán (2011) and Altieri (2012), we understand that agroecology is not a unilinear discourse, but rather an interaction between the codified knowledge of researchers and scientists in dialogue with tacit knowledge from rural and traditional communities. Agroecology isn't a static nor mechanical concept, since it is based on the diversity of the so-called rural and forest social movements in the practical actions and theoretical formulations that are in a constant process of transformation caused by the diversity of the political, social, and cultural characteristics of each community.

---

[13]   Francisco Roberto Caporal, through his books and articles, has contributed to the preparation of this work. He graduated with a PhD from the University of Córdoba – Spain (1998), in the Doctorate course in Agroecology, Peasantry and History, from the Institute of Sociology and Peasant Studies (ISEC), coordinated by Sevilla Guzmán.

A virtue of this diversity of experiences, the interaction, the dialogue and to some extent the conflict between traditional knowledge and scientific-technical knowledge, among extensionist researchers, rural and forest social movements is essential for establishing the epistemological foundation and practices to support the agroecological experience in Latin America (Novaes, 2012).

The complex diversity that composes the populations of Latin America, as well as the history of resistance and the fight against the dispossession imposed by dependent capitalism and an exclusionary *modernization* in the countryside, made it possible that the participatory discussion on agroecology expands in practice as well as theoretically. Dozens of organizations, particularly those formed by rural workers, indigenous communities, and communities of the forest, amplified the discussion and reinforced agricultural alternatives towards an agroecological transition.[14]

Among these organization is MST which since the year 2000 has taken up agroecology as a strategic production framework for the settlement areas and the camps under its influence. This position is reinforced in 2001, when the movement released the booklet *Constructing the Path*, in which it establishes the requirement that "the settlers qualify and master agroecological principles and practices, seeking to build a new model of production that helps us build a new social being" (MST, 2001, p. 90). Although the year 2000 represents a milestone for the insertion of agroecology within MST, since the 1980s,[15] the discussion had already existed among the activists of the movement regarding the need for an alternative framework of socioproductive organization for the Landless.

After its first achievements, MST began to look for alternatives to boost the production of the families and train people with a different vision than that

---

14 La Via Campesina (International Peasants' Movement) stands out as an international movement made up of around 164 organizations in 73 countries in Africa, Asia, Europe and America. In total, it represents around 200 million people, including peasants, small and medium-sized producers, landless peoples, indigenous people, migrants and agricultural workers from all over the world. It is an autonomous, pluralist and multicultural movement with no political or economic affiliations of any kind. For more information visit: http://viacampesina.org/es/.

15 Agroecology was adopted as a production framework by MST at its 4th National Congress in 2000. However, Guhur (2010), Mohr (2014), Borsatto and Carmo (2014) point out that in the Training Notebook No. 10 (MST, 1986, p. 25–28) there is a chapter entitled "the use of alternative technologies", addressing the dominance of multinational corporations over the Green Revolution technological package and the need to build alternatives to the dependent and degrading model of the hegemonic model.

of the man/environment relationship in the areas of the settlements. This is how it developed the guidelines for the Cooperative System of Settlers (SCA) and the Agricultural Production Cooperative (CPAS) in the early 1990s. Even as MST moves forward with the discussions and actions on the social perspective of the cooperative, it has faced several obstacles due to existing contradictions between the conceptions of management of the collective cooperatives and the conception of *cooperation* within the competitive logic of the capitalist market, which has led to debt and the collapse of many cooperatives, starting with the first attacks from FHC's administration.

On the difficulties faced by MST in the 1990s, we can name, in addition to the external particularities, low levels of technical training and the lack of knowledge on the cooperative and the new proposed forms of production. According to Borsatto and Carmo (2014, p. 658), the theoretical conceptions that guided MST

> is based on the orthodox interpretations of the writings of Marx, Kautsky, and Lenin, as well as the Soviet Union and Cuban experiences of agricultural collectivization which in its majority have not shown to be satisfactory in the reality of the Brazilian settlements. This, together with other factors, left political space for the emergence of a new discourse, in which rural knowledge and the environmental question gained prominence, as a result, agroecological discourse emerged.

In the midst of this, Guhur (2010) points out that MST is a movement "of its time" because it is faced with new demands and fights that have grown in the last few years, such as the environmental question, openly confronting the limits and contradictions of the alternatives that they propose to overcome the challenges. It is with this confrontation that MST's IV National Congress deliberated the environmental question (environment, diversity, fresh water, and the Amazons) as a rallying symbol of struggle, around what was known as the Popular Project. In the text *Reaffirmed Political Lines in MST's IV National Congress* (MST, 2000), the model of hegemonic agriculture based on "technological transference, with the use of transgenic seeds, pesticides, exportation of commodities and the monopoly on the land usage by multinational cooperation" is presented as a practice that should be combated.

The document *Our Commitment to the Land and to Life* was also presented at the IV Congress, composed of ten points, among which we highlight, amongst others, "avoid monocultures and the use of pesticides" (Morisawa, 2001, p. 238). Guhur (2010) emphasizes that such a position, demanded a reformulation in the proposals for production, as well as, in the organization of the movement itself. After a period of crisis, triggered by MST's own internal limits and by the actions of the federal government that affected the Movement, the SCA ended up being dissolved and in its place the Sector of Production, Cooperation and Environment (SPCMA) was created.

The environmental question becomes fundamental in the movement's discussions, and agroecology begins to be a strategic production alternative in the proposal of a popular project. Borsatto and Carmo (2014) describe that in MST's Agrarian Reform Proposal in 1995 (MST, 2005), it is already possible to identify the preparation of proposals for the construction of a new production model for the settlers.

Borsatto and Carmo (2014, p. 658) describe that for the preparation of this new proposal:

> The work of Chayanov fundamentally contributes to the shaping the theoretical framework of agroecology (Caporal and Costabeber, 2004). From the chaynoviana conception are drawn concepts on which the methodological proposal of agroecology are based on, such as the farmer, seen no only as a mere object of analysis, but as a subject creating his own existence; a notion of peasant moral economy; a bottom-up approach for drawing up development proposals; the use of multidisciplinary social agronomy analyzes; the economic non-capitalist logic of the peasants; the comprehension of work-consumption balance; the concept of the degree of self-exploitation; the subjectivity of the peasants in decision-making and the concept of differentials optimums.[16]

MST's reorientation was due, among others things, to the following factors: a) the neoliberal reform of the Brazilian state that put an end to sectoral politics, minimum prices, and opened the market; b) the end of Special Credit Programs for Agrarian Reform (Procera); c) the establishment of Via Campesina (The International Peasants Movement). "The first two factors made it difficult to continue the production strategies which until that point were developed by the movement, while the third broadened the array of MST's institutional relationships" (Picolotto: Piccin, 2008 *apud* Borsatto; Carmo, 2014, p. 656).

With this reorientation, the rural workers are no longer mere objects of mobilization in a revolutionary mass and become historical subjects, with knowledge and moral values considered essential for the construction of

---

16 For a broader understanding of Chayanov's work, see the book Chayanov and the Peasantry, organized by Horácio Martins de Carvalho and published by Expressão Popular in 2014.

a fairer, more sustainable, and better society. For this reason, ATER's methodologies begin to value peasant knowledge which is incorporated into the Movement's training processes (Toná; Guhur, 212; Borsatto; Carmo, 2014).

When agroecology is adopted by MST, in addition to referring to a production framework with less degradation to the environment and recognition of traditional knowledge, it involves intense questioning and confrontation with the agricultural policies and technology adopted by agribusiness, which are heavily mechanized, export-oriented and dependent on oligopolized agro-industrial complexes, that do not contribute to the advancement in the fight for agrarian reform (Borsatto; Carmo, 2014).

> the construction of agroecology in MST, implies another conception that is different from all the other currents for the first time a movement of the masses takes up agroecology and makes a component for a political platform for societal change and agricultural model out of it.
> MST, 2005, p. 2 *apud* GUHUR, 2010, p. 144

MST considers that agroecology is one of the paths to combating the new configurations of capitalism in the countryside, delineated by agribusiness. At the closing ceremonies for the 2nd Agroecology Conference of Paraná in 2003, MST supported a protest against the center for research and production of transgenic soy and corn seeds from the transnational company Monsanto, located in the rural area of the municipality Ponta Grossa.

> The area was then occupied by Landless families from the settlements of the region, and converted into the Chico Mendes Center for Agroecology, for a period of 18 months (timeframe after which the families were expelled), with various experimental activities, seed production, and training in agroecology. According to Gonçalves (2008), this fact shook up the relations among the organizing entities of the Congress, causing some of them to pull out because they did not support the anti-capitalist nature that the event had adopted, and also because they felt discredited in the organization. It was an important political moment, since although genetically modified crops had been expanding in the country clandestinely, there still wasn't a definitive decision from the Federal Government on the matter. The occupation of the multinational company Syngenta Seeds, also in Paraná, and the seedling nursery Aracruz Celulose, in Rio Grande do Sul, in 2006, followed in the same vein.
> GUHUR, 2010, p. 145

In Brazil's new shaping of land exploitation, agribusiness is a hegemonic model, preserving fundamental elements from the latifundia and consolidating an alliance among financial capital, banks, large landowners, and the transnational corporations that control inputs, prices, the trade of goods, the bourgeois media, and the state apparatus.

The changes imposed by agribusiness, since the 1990s, has led to a restructuring of rural exploitation. Therefore, in MST's reorientation, there is a significant change in the nature of the fight for land and for rural structural changes, it is no longer about confronting the latifundia and its aggressive methods but proposing a new project for the countryside, a project that has been named *Popular Agrarian Reform*.

> This proposal for agrarian reform reflects part of the anxieties of the Brazilian working class in constructing a new, egalitarian, united-in-solidarity, humanist, and ecologically sustainable society. Thus, the proposals for necessary measures should be part of a larger process of change in society and, fundamentally, for the alteration of the current structure of the organization of production and the relationship of the human being with nature, in a manner that the entire process of organization and development of rural production aims to overcome exploitation, political domination, ideological alienation, and destruction of nature. This means valuing and guaranteeing people's work as a condition for human emancipation and the construction of dignity and equality among all and the establishment of harmonious relations between human beings and nature.
>
> MST, 2013, p. 149

For the proposal of the *Popular Agrarian Reform* agroecology is a technological framework adopted as an alternative for the socioproductive organization of the settler and encamped families, because it represents a means of increasing the productivity of labor and areas, in harmony with nature, with the possibility of confronting and combating agribusiness and private and intellectual property coming from the registration of patents on seeds, animals, natural resources, and biodiversity (MST, 2013).

For Gonçalves (2008), what mobilizes MST is the refusal of the standard of agricultural development available in the country, emphasizing the need for preserving and reconstructing peasant agriculture by means of agrarian reform, in addition to proposing forms of peasantry management and participation in cooperative and agroecological systems of production.

Toná and Guhur (2012) observed that a more recent and broader conception of agroecology is found in the making, which has the rural popular social movements as its political cornerstone. This branch doesn't see agroecology only as a mere technological solution to the structural and temporary crises of the economic and agricultural model. Agroecology, as observed by Via Campesina and MST, is understood as part of a strategy of struggle and combating agribusiness, the exploitation of workers, and the degradation of nature. In this conception, agroecology includes the care and defense of life, food production, political and organizational consciousness (Toná; Guhur, 2012, p. 66).

MST considers that the change in social, ecological, and, above all political, and technical rationality of the families helps to overcome the new dynamic of rural capitalism, based on extremely severe relationships of domination, such as the presence of genetically modified seeds and the connections among the transnational agribusiness capitals (chemical, food, and finance) (Gonçalves, 2008). Despite the emphasis that the program Popular Agrarian Reform gives to agroecology, Luzzi (2007, p. 130) describes that the incorporation of this production framework

> for the settlers it is not a simple question, it involves many factors and the changes don't always happened at the speed desired. Espousal of the theme by the leaders of MST occurs in a more accelerated manner than that which has been occurring in the settlements, in practice by the settlers. Although MST is investing heavily in training and capacity-building in agroecology, the change is still rather slow. The modernizing ideology continues to exercise a strong power of influence on the settlers and, why not say it, on various leaders.

However, even though MST doesn't have the power to make the radical transition to agroecology, it is strong enough to go into battle with agribusiness, in particular, having permanent campaigns against the use of pesticides and defending that seeds, instead of being a monopoly for a few corporations,[17] should be a patrimony of the people in service to humanity.

---

17 3) described that biotechnology and transgenics, as they have been used in agricultural production, are developed on the basis of reductionist techniques that promote monocultures and produce severe genetic and laminar erosion. They point out that in addition to standardizing the production of plant foods on 15 species that account for 90% of the food produced, based on four crops (wheat, rice, corn and soy), which account for 70% of the world's production and consumption, these are procedures that eliminate biological diversity, preventing the natural genetic improvement of populations.

The permanent campaign against the use of pesticides, besides questioning the evils of the use of the chemical defenses, be it for human health (with an uncountable number of registered cases of contamination, just as many of them workers as consumers), or be it for the pollution and deprivation of natural resources, requires the adaptation of the production system to a cleaner basis, linked to the principles of agroecology.[18]

In this search for the democratization and non-commercialization of seeds, as well as the fight against the use of pesticides, we highlight the actions carried out by women that make up Via Campesina. Pinassi and Manfort (2012) present a paper with diverse activities of women from Via Campesina that seek to denounce the harmful effects of food consumption produced on the basis of genetically modified seeds and on the use of pesticides.

The protagonism that the women have been taking on in the restructuring of the socioproductive organization for agroecology, is just as important as the actions of confronting the patriarchy in the internal structures of the organization of the working class. "These women demand, that we think urgently about a radical alternative to the system, an alternative that is constructed on the realm of freedom and substantive equality"[19] (PINASSI; MAFORT, 2012, p. 155).

We can hypothesize that the fight for agroecology is related to what Mészáros (2011) calls substantive equality. If capital promotes formal equality, than the anti-capitalist social movements are fighting to build substantive equality in terms of gender, ethnicity, generation, and above all, overcoming class exploitation. It isn't by mere chance that the women of MST organize struggles for economic independence, not-subordinated by the husband, while at the same time getting involved in class, gender, and environment issues, in an interesting intertwining (Pinassi; Mafort, 2012).

Thus, the role of the woman in MST contributes to the progress of the discussion on agroecology, their actions adding to up with other farmers, extensionist

---

18   Silvio Tendler's documentaries O Veneno Está na Mesa 1 and 2 (The Poison is on the dinner table 1 and 2) are a fine critique of the Green Revolution. In the first film, the structures and contradictions of the conventional "Green Revolution" model are described, as well as the basis of transgenic seeds and the need to use pesticides. This production model puts 5.4 liters of pesticides on the dinner table of every Brazilian. The second section presents experiences of agro-ecological production as an alternative to the contaminating model, and also presents some advances in relation to public policies. However, it draws attention to the challenges posed by the corporations that have monopolized the food production chain.

19   On the debate about subjectivity and gender in MST see: Silva, Crístiani Bareta de. Men and Women on the Move: Gender Relations and Subjectivities in MST. Florianópolis: Momento atual, 2004.

technicians, and even consumers who together compose a significant portion of the citizens that stand up in defense of agroecological production, the example of National Coordination for Agroecology (ANA) and Brazilian Association of Agroecology (ABA).

In the realm of action, within the state, the following stand out: the National Policy for Agroecology and Organic Production (PNAPO), the Program for the Acquisition of Food (PAA), and the National School Nutrition Program (PNAE), which more or less within the limits of the capitalist state have been recognizing the need to develop organic and agroecological practices.

However, in the midst of the complexity and dispute over the agroecological framework, we cannot ignore that there are many organizations that follow an eco-developmentalist manual from international organizations such as the IMF and World Bank, connecting to the area in an opportunistic and/or reformist manner, with the goal of developing green markets with discourses on sustainability and valuing the product. This fact is illustrative that there are, at least, two branches linked to agroecology, one related to green markets, derived by capitalist logic and the other related to MST campaigns, that visibly don't dissociate the structural bases of the production of a reflection on social issues such as: young peasants, gender, class struggle, among others.

> In this context, agroecology is not restricted to the development of the ecologically-based farmers' experiences, highlighting processes of social organization that are oriented on political struggle and social transformation, going beyond the immediate economic and corporate fight and the local, and at times assistance-oriented actions with the farmers. In fact, agroecology possesses a specificity that references the construction of another rural project. In the meantime, such a rural project is incompatible with the capitalist system and depends, ultimately, on overcoming it.
>
> TONÁ; GUHUR, 2012, p. 63

The fact that agroecology is constructed and discussed in dialogue with a variety of actors has brought critical perspectives of knowledge and new strategies to the mediation of knowledge, for instance of social technology (TS), which contributes, according to Caldart et al. (2002), Kolling, Nery; Molina (1999) and Almeida; Antonio; Zanella (2008) to boosting rural education. TS, when questioning the myth of science's neutrality and technological determinism, seeks to deconstruct the belief in expert solutions and places technology as a collective construction with and by its participants, opening up the possibility of generating sociotechnical solutions from the lived social relationships (Fonseca, 2009).

In relation to rural education Caldart (2009, p. 44) describes that

> In the affirmation of the importance of the democratization of knowledge for the access of the working class to historically accumulated or produced knowledge through class struggle, education in the countryside brings a more radical problematization of its own production of knowledge, as a critique of the myth of: modern science, cognitivism, and foolish bourgeois rationality; as a demand for a more organic connection between knowledge and values, and between knowledge and the totality of the training process. The democratization demanded, however, is not only about access but also about the production of knowledge, involving other forms of production logic and overcoming the hierarchical vision of its own knowledge from capitalist modernity. The questions today about the construction of a new agricultural project/model, for example, do not only involve rural workers' access to existing science and technology. This is exactly why they are not neutral. They were generated from a certain logic, and that is for the reproduction of capital and not the worker. This science and these technologies should not be ignored, but need to be overcome, which requires another logical thinking, and production of knowledge.

Enio Guterres (2006) explains that agroecology in Brazil develops in a restrictive manner, or rather doesn't develop because the majority of the teaching institutions approach the agroecological question in a restrictive manner. The author also highlights that there isn't enough technical assistance to follow along with all of the people that begin the agroecological transition process.

It is important to emphasize that, between the twentieth and twenty-first century, the agroecology courses began to formally emerge on the national scene. By the end of 2013, 136 curses were identified to be in operation, 108 of which were on the technical level, 24 on the higher-education level, and 4 on the post-graduate level, the vast majority of these courses, 44 of them, were located in the northeast region of the country (Balla, 2014). Considering the existence of many conceptions on agroecology and keeping in mind that the option for one or the other dimension depends on the arrangements of the diversity of the territory where it is going to be developed, the next topic presents some reflections on the many dimension of agroecology.

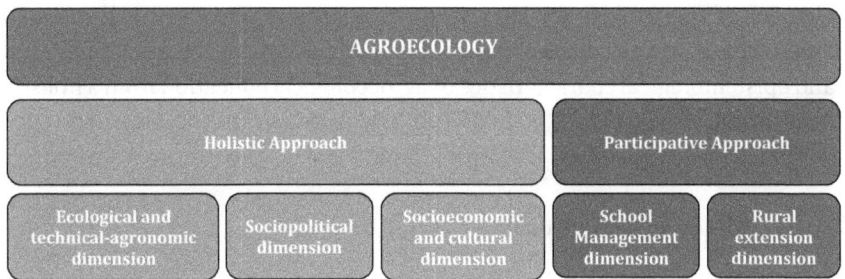

FIGURE 5.1 Approaches and dimensions of agroecology
SOURCE BY THE AUTHORS

6   The Dimensions of Agroecology

Agroecology follows a critical branch of the Green Revolution and an alternative proposal to theories and methodologies that try to reconstruct the framework of socioproductive organization in agriculture, that can just as well manifest itself in the design of the green markets, as it can manifest itself in the critique of agribusiness, as we have explained in the course of the text.

Bearing in mind that we consider that the role of agroecology in Brazilian society must go beyond the design of the green markets, we will now discuss the theoretical and practical field of knowledge. For this purpose, we center our analyses on the conception that agroecology is a field of knowledge that connects with historical processes, economics, and culture that structure the organization of agricultural production within Brazilian society, with nuances between scientific and popular knowledge in the organizational framework of the agricultural workers' training from MST.

Due to this position, we utilize the categories holistic and participatory to characterize that agroecology is a scientific-technological framework that is not disengaged from the other dimensions of human life, it is an educational activity[20] of men by men, in conjunction with nature and society, as shown in Figure 5.1.

With this premise in mind, we are dedicated to investigate in which aspects the curriculum model for the technical course in agroecology in MST approaches the holistic methodology, given that the historicity of the movement around a collective project of social organization constitutes a central

---

20   Among the works that seek to theorize agroecology from an educational perspective, we highlight: Guhur (2010), Caporal and Azevedo [orgs.] (2011), Lima et al. (2012) and Mohr (2014).

element of the struggle for land, in defense of nature, society, and man. To argue the fact, we will discuss in the subheading, following the methodological and epistemological characteristics of agroecological education from a holistic conception.

7 The Holistic Approach

The holistic concept was coined in the midst of a critique on the discourse of rationality, objectivity, and quantification as a unilinear means of arriving at knowledge. According to Tristão (2004, p. 131), the holistic approach is configured as a resistance to the non-integrated and reductionist approaches that cannot "see the integrative activities between the live systems and the social systems, and their interactions with the environment".

Brian Swimme (s/d) quoted by Teixeira (1996) explains that the holistic approach overcomes scientific reductionism because it is based on the following principles:
– Interaction: the identity and the existence of the elements that compose the universe only have meaning within the relationships that they establish between each other, in a dynamism of energy;
– Participation: our knowledge only attains a qualitative consciousness, when it is placed in interaction with other sets of knowledge;
– Analysis and synthesis: the production and the interpretation of the world depends on understanding the origin and the purpose of the knowledge;
– Self-organization: the universe is only understandable from an analysis of the totality and complexity of knowledge;

With a foundation in these principles, we can affirm that the holistic approach proposes a systemic, complex, and transdisciplinary vision of knowledge that tries to overcome the atomistic vision of the newtonian-cartesian paradigm (Teixeira, 1996) The systemic, complex, and transdisciplinary vision understands that not everything is constructed only on the sum of its parts, but on the interdependence among the phenomena that comprise the historical context where the *coevolution* of human life takes place.

The transdisciplinary approach establishes the dialogue between the more varied forms of scientific knowledge and wisdom constructed by the rural workers, in the case of agroecology. Crema cited by Teixeira (1996, p. 287) highlights that the professionals that work on the ethos of a holistic approach should be scientists, researchers and philosophers with a willingness to have "inclusive" experiences.

Santomé (1998) describes transdisciplinarity as a form of establishing dialogues between the methodological approaches and epistemological concepts of many academic disciplines in that the limits between them transcend, disappearing so that an explicative and totalizing system of human life emerges, contributing to the birth of a *macrodiscipline* capable of serving as the means of interpreting and intervening in reality in an all-comprehensive manner. Transdisciplinarity, especially, with regards to the concept of complexity, maintains relations with the orientations that Moisey Pistrak presented to the professors of the Soviet Union's educational system, by means of a methodological charter, in which the educator explains that

> complex means the concrete complexity of the phenomena, taken from reality and unified around a certain theme or ideal. The concrete complexity of the phenomena grasped from reality refers to life, and this is the issue of work: 'From this point of view, work is the foundation of life for people. It follows that people's work activity is the center of the study'.
> FREITAS, 2013, p. 35

For Luiz Freitas (2013, p. 35), the relationship among nature, work, and society is indissociable, they should be studied in a connected manner because it is only in this manner that it reflects the complexity of reality, "its dialectic and currentness [...] its contradictions and struggles – its development in nature and in society, based on people's labor".

In this regard, the systemic, complex, and transdisciplinary thought are the means of characterizing the holistic approach, however, this approach is not restricted to these concepts. For Capra (1995, p. 200), "a holistic approach [...] is crucial when one wants to understand how a certain system is immersed in larger systems", because it allows us to comprehend how economic activities, for example, "are immersed in the cyclical processes of nature and in the system of values from a certain culture".

Due to this consideration, we can affirm that the holistic dimension of agroecology is not a simple junction of the areas of agriculture and ecology, but a way of elucidating that these systems connect with the economy, culture and history in a more complex manner, and truly with all of dimensions that involve human production. Therefore, agroecology when developed and studied under the holistic approach, contrary to the compartmentalized or isolated form of analyzing reality, should be mindful of a broader, integrated, and complex analysis, in that farm work is more than the management of techniques.

From this premise of the systemic integration among the different dimensions of human life, Sevilla Guzmán and Otman (2004) affirm that agroecology is constituted on three dimensions: a) ecological and technical-agronomic; b) socioeconomic and cultural; and c) sociopolitical. To explain how these three dimensions contribute to the construction of holistic agroecological theories and practices, we present the meaning of each one of them, in the next part of this book.

a) *The ecological and technical-agronomic dimension:* This brings the ecological question to the center of the debate meaning it redesigns the process of human beings' *coevolution* and their environment. It is not about returning to rudimentary work methods from primitive society but about reverting the damage caused by capitalism. Agroecology cannot lose sight of some essential dimensions when it is presented as an alternative production framework for food production. In this sense, having Machado and Machado Filho (2014) as a reference, we understand that the ecological and technical-agronomical dimension can be broken down into four subdimensions: a) scale; b) environment; c) energetic; and d) technical.

The scale subdimension is about the commitment to supplying clean food in quantity and quality to feed humanity and to be an alternative to agribusiness, "scale has to be planetary that is the scale of agribusiness, which today controls the supply of *food* to the world" (Machado; Machado Filho, 2014, p. 191). The agroecological framework as a strategy of resistance to capitalist agricultural production cannot be disavowed from its obligation of supplying food, beyond the microcosm of the agroecosystem, in other words, of self-consumption, but also supply food for society as a whole.

The environmental subdimension seeks a way to revert the process of *coevolution* between the social systems and the natural systems, aiming to protect the environment in all of its entirety. It is not about neoliberal processes that propose valuation of the environment with an economic aspect (monetary) in a deterministic manner. It is about having the human and natural sciences as a committed scientific-historic process for overcoming the utilitarian vision of the use of natural resources. Thus, the development of labor, management, and utilization techniques gain meaning when they have the intention of reverting the negative environmental impacts of the capitalist model.[21]

---

21   In the case of agricultural management, for example, instead of using synthetic fertilizers and pesticides, through the ecological dimension it is possible to enhance the maintenance and improvement of the level of organic matter (OM) in the soil, OM besides being the catalyst for life in the soil, is the main reservoir of CO_2 on the earth's surface; 1g of OM retains 3.65g of CO_2; if the level of OM in the world's soils increased by 1%, the level of CO_2 in the atmosphere would return to pre-Industrial Revolution levels (34).

The energetic subdimension is handled by Machado and Machado Filho (2014) as a *touchstone* of agroecology because it doesn't associate agroecology with a low demand for energy but with finding alternative sources of energy, as is the case for solar energy. In this sense, the fossil fuel portion that enters the system of production should be less than the energy produced by photosynthesis, at times through routes not directly photosynthetic, such as, for example, organic material that feeds an important part of the life of the soil, therefore, a primary source of this energy is always solar (Machado; Machado Filho, 2014).

The technical subdimension is intimately connected to the question of labor and by extension to teaching and learning; it concerns how to develop the management techniques and utilization of natural resources, without losing sight of the principles of agroecology and the means of how to develop them.

Machado and Machado Filho (2014) present the following management practices that are directly related to the production techniques for agroecological food: a) don't pummel the soil; b) don't plough; c) don't soil grid; d) don't subsoil, or in other words, dispense of any procedure that moves the soil because, otherwise, the pillar, the foundation of agecology (trophobiosis, ethylene cycle in the soil and transmutation of the elements to low energy) is not unleashed and it doesn't happen; e) respect the biodiversity; f) direct planting; g) crop rotation; h) plant companions; i) allelopathy; j) crop association; k) farming/ploughing association l) perifocal control of parasites; m) dead coverage (mulching) n) overseeding; o) green fertilization.

The technical-scientific process of this dimension could not, in any manner, abstract from the processes of political praxis of confronting the conventional technical model, nor from the practices and knowledge of resistance developed by self-governing communities. These techniques make reference to the construction of an epistemological foundation with strong consideration for an ecological variable that can revert the current environmental crisis by means of less degrading metabolic actions.

b) *Socioeconomic and cultural dimension:* Agroecology is also a way of critically positioning oneself in relation to the conventional school curriculum, structured on the logic of productivity and cultural standardization, especially, when it is in regard to technical-professional training. In this perspective, agroecology is a form of resistance to the traditional curriculum and seeks to insert the cultural practices formed by historical individuals silenced by the traditional schools into the structure of conventional teaching.

Thus, the socioeconomic and cultural dimension seeks to include the grievances and demands of class and the organized traditional peoples, which as Norgaard (1994) emphasizes were considered part of nature and in process of *coevolution* with it, which resist and are shown to be an alternative to the *modernization* of agriculture.

It is not exclusively about raising the level of productivity, clearly this is also important, especially to be able to meet the global demand for food. However, it is also about, in equal proportions, perfecting the endogenous potential, making dignified work conditions possible; this is a "knowledge-intensive" proposal as opposed to "capital-intensive" systems (Machado; Machado Filho, 2014, p. 192). The socioeconomic and cultural dimension seeks to emphasize the design and execution of the social processes, of the processes of choosing the tools for socioproductive organization, respecting the particularities represented in the sociocultural manifestations in each agroecosystem.

c) *Sociopolitical dimension*: Agroecology, being a alternative production framework and counter to the *Green Revolution*, is founded on a less degrading and potentially fairer praxis, justified on a dialectic of *coevolution* in many perspectives of the metabolic relationship,[22] including its own procedures for building knowledge. The sociopolitical dimension helps us to explain how the process of training professionals should be organized in order to work in the agroecology sphere.

We consider that the sociopolitical dimension is indispensable to the professional training process, because it represents the possibility of overcoming the unitary and/or individual vision, giving incentive to individuals to get involved in the day-to-day political praxis, indicating that that man's action with nature and with other men take place around a *coevolution* with other lives.[23] Political praxis from the social actors that make use of agroecology should establish political processes that respect and seek to address all of the dimensions involved in the agroecological production process.

---

22 In his definition of the labor process, Marx made the concept of metabolism central to his system of analysis, describing labor as first and foremost a process between human beings and nature, in which human beings, through their own actions, mediate, regulate and control the metabolism between themselves and nature. The human being sets in motion the natural forces of his own body, his arms, legs, head and hands, in order to appropriate the materials of nature so as to adapt them to his own needs.

23 In a narrower sense, social praxis is the activity of social groups or classes that leads to transforming the organization and direction of society, or to making certain changes through the activity of the state. This form of praxis is precisely political activity (Vázquez, 2011, p. 232). As an example, we can cite the National School Feeding Policy (PNAE), the result of a dispute between social movements in the countryside, small farmers and extension technicians over better conditions for transporting produce from the countryside.

The sociopolitical dimension involves an ethical commitment to the other and to the environment. From this dimension, ethics "is a reflection on the attitudes and appropriate actions with respect to beings and the processes of relevance, where relevance has to do with the fact that these beings and processes are important in themselves" (Heyde, 2003 *apud* Caporal; Azevedo, 2011, p. 48). Agroecology doesn't embody incorrect or morally problematic actions, such as: throwing toxic waste in the ocean, polluting the streams or riverbeds with the use of herbicides, emitting gases that cause the increase in global warming, the contamination of the soil and water with chemical residue, the utilization of slave labor, social inequality, etc.

Therefore, ethics in agroecology has a strict connection to the *precautionary principle*, in a way that its development aims to avoid any increase in risks, in addition to those that already exist, in relation to the development and the application of new processes and technologies (Caporal; Azevedo, 2011). The discussion and concrete actions regarding gender are also a part of the sociopolitical dimension in agroecology which should be adopted as a means of struggle and political, social, and economic emancipation for the individuals in the face of the advances of capitalist forces. The question of gender, in addition to being the gateway for women in the sociopolitical scene, represents the possibility of breaking away from production and reproductive labor that sustain patriarchal capitalism, rethinking not only women's position, but also the material structure of conventional agricultural production and its replacement by alternative agroecosystems.

The historical cause of women's emancipation can not be achieved without affirming the demand for *substantive equality* which directly challenges the authority of capital, prevalent in the far-reaching macrocosm of society and equality in the microcosm of the nuclear family (Mészáros, 2011).

Because of this, the gender question includes the most varied oppressed categories that fight and resist society centered on exploitation and oppression. Therefore, we suggest the need to overcome the conceptualization of gender beyond the simple differentiation between man and woman, and embrace the discussion that include overcoming all of the types of oppression, starting with the relationship between man and woman in the nuclear family, but also of other categories that are and were oppressed in the history of the development of capitalist society.

In short, the three dimensions (ecological and technical environmental, socioeconomic and cultural, and sociopolitical) which constitute the base for the holistic approach aim to represent an epistemological base for agroecology that transcends the sum of ecology and agriculture.

8 The Participative Approach

Agroecology as a scientific and technological framework should be careful to ensure that its solutions for the crisis are not based on the same epistemic postulates from the scientific paradigms that caused the problem in the first place[24] (Gomes, 2011). In considewration, it was proposed that the epistemological bases for agroecology develop in the sense of critical theory of knowledge where both scientific knowledge and day-to-day wisdom, like the wisdom of rural workers, also known as traditional, local, or indigenous knowledge, are considered in *coevolution* (Gomes, 2011).

The principles of agroecology recognize that there is a technical-scientific potential in the local communities that is already known and is capable of boosting substantial change in agroecosystems. Therefore, the activities of teaching, research, scientific and technological application, should be developed with a foundation in a perspective that ensures the reorientation of theoretical and practical labor. Agroecology, as an alternative framework, should be formulated from a critical perspective and from resistance to the prescription

---

24   João Carlos Gomes (2011), when dealing with the epistemological bases of agroecology, presents a strong concern for agroecology when it is affirmed as a new paradigm. He points out that agroecology is an alternative to overcome the crisis imposed by the Green Revolution paradigm. It must avoid using the same tools as the Green Revolution and, in a broader sense, science in general. Following the historical path of the philosophy of science, it analyzes British empiricism with Francis Bacon (1561–1626), the rationalism of Descartes (1596–1650), the positivism of Comte (1798–1857), the neopositivism of the Vienna Circle, the critical rationalism of Karl Popper (1902–1994), arriving at the new philosophy of science and contemporary debates on science, defending the context of research and plurality in science, a new alliance between human beings and nature. He also defends the characteristics of a new paradigm for the theory of knowledge, encompassing the foundations of ecology, natural epistemology, evolution, politics and participation, arriving at epistemological pluralism in Agroecology. Thus, he makes a brief map "through the critical reconstruction of the theoretical conceptions of scientific and technical knowledge, allowing a reflection on the progress of modern science, drawing attention so that the search for the epistemological bases of agroecology does not follow a mistaken path" that demonstrates the vicious cycle generated in the contemporary crisis. Concerned with the ideological deviations imposed by the logic of capital, Mészáros (2011, p. 243), in a passage on pluralism and legitimation, draws attention to the very nature of capital as constituted by an irremovable plurality of capitals "which, ideologically heralded, radically excludes the legitimacy of a objection made from the point of view of the alternative and structurally subordinate hegemonic class, in this sense, we have to be careful when bringing the issue of pluralism into the debate, not to fall into a 'false pluralism' whose real class substance is revealed in major crises".

for scientific and technological transfer, in which the experts are seen as producers of knowledge and the workers as deprived of knowledge.

The participative approach in agroecology is applied to the joint actions of researchers/extensionists side by side with workers and traditional communities in resistance, calling for a dialogue between the parties involved in the day-to-day of the agroecosystem to be at the center of the debate. With the metabolic relationship between human beings and nature at its core, the participative action should overcome the perspective of objectification of the actors (peasant workers, native communities, researchers, extensionists, technicians, and the environment) including them as subjects of its historical process.

According to Minayo (2008, p. 163), the development of participative methodological procedures emerged in the 1960s, under the influence of critical thought with regard to the social reality in Latin America, with the goal of combining investigation, participation and politics. The participative approach is substantiated on the following conceptions: a) the idea of a popular subject; b) the idea of a political project backed by a popular front; c) the concession of local space as a political locus; d) the role of the researcher as a transformative political actor.[25] The methodological procedures of participation aim to guide the peasants and the most relegated social groups of society to insert themselves in the social processes and to integrate themselves into the political debate, in order to find solutions to the concrete problems of life in society.

In the field of agrarian sciences, the premises for developing participative actions are based on the interpretation of the theory of agrarian systems.[26] Moreira (2003) highlights that the theory of agrarian systems was constituted as a means of resolving the problems experienced in agroecosystems. However, the author affirms that this theory has some limitation, such as: a) the limited progress in the independence of the farmers in relationship to

---

25 Minayo (2008) points out that these principles must be accompanied by the following assumptions: a) social inclusion can only be achieved if the economically and socially excluded sectors become aware of their interests, organizational practices and real social and political significance; b) social research can be a powerful vehicle for these changes.

26 In the 1970s, a certain contingent of researchers began to question the fact that many "laboratory scientific truths" were not really "true" in the field, because the controlled conditions of the experimental stations were not capable of reproducing the physical, natural, socio-economic and cultural contexts in which the technologies were applied, thus not obtaining the same results. These scientists began to listen more to farmers and to carry out research on farms, which led to the emergence of the current Farming Systems, mainly among the English and French (On Farm Research, Farming Systems Research, On Farm Client Oriented Research and Farmer Participatory Research) (Moreira, 2003, p. 22).

transnational companies and to fossil fuels; b) flaws in materializing agroecological processes based on *coevolution*; c) little progress has been made in the implementation of transdisciplinary practices; d) prevalence of the verticalization of knowledge between researchers and farmers, thus lacking a horizontal subject-subject relationship which is characteristic of the Participatory Action Research movement; e) its approach has a strong adherence to the philosophical premises of conventional science (Sevilla Guzmán; Woodgate, 2013).

In addition to the technical and methodological limits of the theory of the agrarian systems that involve participation, we cannot forget to bring attention to the limits imposed by ATER's policies, pre-established by the state, which legitimize the relationship among teaching, research and extension from the perspective of the *modernization* of agriculture, based on the logic of transference of knowledge.

Participatory action, when applied to the field of agroecology, on matters of principle and practice, is aligned with the resistance struggle against capital's socioproductive utilitarian model and seeks a real integration between so-called scientific knowledge and local and indigenous knowledge, resulting in solutions from its endogenous potential for the design of agroecosystems (Gomes, 2011).

Gomes (2011, p. 34–35), with a foundation in Campos (1990) affirms that the methodologies of participation: a) overcome the opposition between scientific and traditional knowledge, allowing for the integration of knowledge in a non-subordinated manner; b) integrate scientific knowledge with that of popular knowledge, aiming to overcome the methodological, theoretical and technical issues, brought on by the excessive mediation of scientific rationality; c) involve an intrinsic relation between the epistemological realm and the practical realm.

Therefore, the mechanism of participation, from the perspective of agroecology, according to Casado; Sevilla Guzmán; Molina (2000) quoting Robert Chambers, should: a) revalue popular knowledge from the farmer, local or indigenous; b) develop principles of Participatory Action Research in agricultural research; c) critique the rural development undertaken by international development organizations; d) develop participatory farming technologies, moving epistemologically and methodologically closer towards agroecology (Robert Chambers *apud* Casado; Sevilla Guzmán; Molina, 2000).

These conditions advocate for an integration of knowledge in a non-subordinated manner. However, it is not about ignoring the scientific and technologic growth that has developed throughout history, and to a lesser extent, is it about subordinating indigenous knowledge or the actions of knowledge transference from specialists. It is about confronting them in the attempts to overcome the methodological, theoretical, and technical problems caused by the rational scientific mediation that normally tends to filter and adapt other knowledge to its schemes, impoverishing it.

In agroecology, scientific activity is not an independent activity nor is it above reproach because every "epistemological construction is the result of a sociocultural situation of historical nature" and scientific production cannot guarantee "the separation between reason and passion". Therefore, agroecological science is the result of the intellectual and political context and praxis of those who produce it, from where they produce it (Casado; Sevilla Guzmán; Molina, 2000, p. 155).

In this perspective, agroecology, by means of participative interaction, is based on the conception that it is not enough to regard scientific knowledge as closed-off and isolated axioms because both the researchers as well as the extensionists and the rural farmers possess knowledge about agricultural production that should be shared, in the direction of realizing a holistic approach to agroecology.

9   Educational Dimension

In addition to the ecological management of natural resources, agroecology needs to be thought of along the lines of a holistic and complex perspective, so that it can contribute and redirect the course of *coevolution* between nature and society in its multiple inter-relations in the construction of reality and knowledge.

The training and educational process regarding the principles of agroecology are organized in a manner to provide an alternative for the conception of how human beings modify and transform the world. In this light, it a) should rethink and propose new techniques for human beings to relate metabolically with nature, transforming it in the sense of meeting its needs through labor; b) should encourage political praxis as a form of social insertion in the planning, decision, and execution processes in the ecosystems; c) should not fall into reductionism and utilitarianism from the educational model disseminated by the *theory of human capital* and *competencies* in society.

The education model offered by the Brazilian capitalist state, in addition to supplying fragmented and reductionist education, works with the impetus of preparing for the labor market, without going in its perverse duality, which can be synthetically pointed out between: a) private education revolving around the bourgeoisie and the middle class, with the majority of graduates of this education in higher educational courses; b) public education, supplied precariously by the state, with the goal of preparing the labor force for the market and/or to building the reserve forces.[27] Opposing this proposal, we consider that the educational dimension, on the principles of agroecology, is conceived from an ontological or ontocreative perspective mediated by the history of conscious activity from human beings through labor, which is not restricted to work or employment, but to the totality of human life in *coevolution* with nature (Frigotto, 2010).

According to Caporal and Azevedo (2011), agroecology is a science for the sustainable future, which, in addition to considering knowledge and native work relations, aims to integrate and connect knowledge and techniques concerning different scientific disciplines.[28] As Casado, Sevilla Guzmán and Molina (2013) described, it is about approaching knowledge in a pluralistic manner, in which the limits and decisions of authority are democratically established and integrated.

> Unlike the compartmentalized forms of seeing and studying reality, or the isolationist methods of conventional science, based on the cartesian paradigm, agroecology integrates and connects knowledge from different sciences, such as popular knowledge, allowing as much for the comprehension, analysis and critique of the current model of development of industrial agriculture, as the design for new strategies for rural development and styles of sustainable agriculture, from a transdisciplinary and holistic approach.
>
> CAPORAL; AZEVEDO, 2011, p. 51

In this light, transdisciplinarity can be understood, just as Santomé (1998) describes, like a transcendent concept in which the lines between the many disciplines disappear and a total system is constructed that exceeds the level

---

27 The process of development of Brazilian dependent capitalism, and consequently the institutionalization of education more or less, was based on a process of education aimed at meeting market demands, the broad need to adapt the workforce in exodus to the new reality and the absence of endogenous knowledge production (Xavier, 2008).

28 Jurjo Santomé, while falling into a post-modern outlook, makes some contributions to our discussion. When analyzing the structure of curricula in disciplines, he describes that disciplinary knowledge refers to a set of abstract structures and intrinsic laws that allow particular classifications of concepts, problems, data and verification procedures according to assumed models of coherence. In most cases, this is how the various disciplines are formed, with concrete names intended to represent the different parts of human experience and knowledge (Santomé, 1998, p. 103).

of reductionist relations, the integration should happen within a *omnicomprehensive* system.

In the pursuit of common goals and an ideal of epistemological and cultural unification, cooperation is such that it can be considered the birth of a more complex formation. To that end, work plays a fundamental role in the educational dimension, not merely as methodology or didactic techniques for the learning process, but as a ethical-political principle, thus, it is simultaneously a duty and a right.

> An duty because of the fact that it is fair that everyone collaborates in the production of the material, cultural and symbolic goods fundamental to the production of human life. A right due to the fact that the human being is a being of nature who needs to establish, through conscious action, a metabolism with the natural environment, transforming it into goods, for its production and reproduction.
> FRIGOTTO, 2010, p. 61

In short, knowledge building from agroecosystems and an interdependent *coevolution* between the social being and the environment, agroecology must propose guidelines for training that seeks to establish the process of management and transformation of nature, in order to overcome the utilitarian and reductionist logic of the model of capitalist agricultural production.

CHAPTER 6

Transnational Corporations, the MST's Agroecological Agenda and Agroecology Schools

1   Introduction

> The "Green Revolution" in agriculture should have solved the problem of hunger and malnutrition once and for all. Quite contrastingly, it created monster corporations, such as Monsanto, which established their power all over the world in such a way that it will take a great agenda of popular action aimed at the roots of the problem to eradicate it. Despite this, however, the ideology of strictly technological solutions continues to be propagated to this day, despite all its failures.
>
> ISTVÁN MÉSZÁROS, The power of ideology

As we started writing this chapter, we were "surprised" by two news stories: a) Bayer's merger with Monsanto, in which two "monster corporations" turned into a much larger monster; and, b) the UN announced that South Sudan had entered the food insecurity map.[1] According to another UN report, 5 billionaires hold wealth equivalent to half of the world's population and about 1 billion people go hungry every day (Ziegler, 2013). Apparently, there is no prospect of improvement. On the contrary, studies in the area point to an increase

---

1   "The UN reported on Monday (20) that more than 100 thousand people are suffering from hunger in South Sudan and that about 1 million are on the verge of food insecurity in the country. Hunger has become a tragic reality in parts of South Sudan, and our worst fears have been realized", said Serge Tissot, FAO representative in the country, in a press release issued in conjunction with the United Nations Children's Fund (UNICEF) and the World Food Program (WFP).

"Many families have exhausted all the means they have to survive", he added, explaining that these people are mostly farmers who have lost their livestock and even their farming tools. The situation is the worst famine since fighting broke out over three years ago between rival forces – the Sudan People's Liberation Army (SPLA); parties loyal to President Salva Kiir; and the Sudan People's Liberation Army in Opposition (SPLA-IO). According to the FAO, UNICEF, and the WFP, "4.9 million people – more than 40% of the population of South Sudan – need urgent food assistance and need help to grow plants" (UN, 2017).

in poverty and income concentration and the intensification of catastrophes at unprecedented levels.

"Monster corporations" are increasingly free to advance an agenda of commodification of life, the takeover of territories, and the free circulation of their shares on the stock exchanges. UN reports refer to "poverty management" and no longer to "overcoming poverty" or "development strategies". With the advance of neoliberal policies based on the direct and indirect privatization of public services, that is, a minimum state for workers and a maximum state for financial capital, barbarism will only gain more fuel: we will have the emergence of more and more slums, an increase in unemployment and underemployment rates, destruction of the lives of civil servants, low-intensity wars, and the resurgence of fascism, among others.[2]

In Brazil, we recently had a parliamentary coup, more complex and difficult to understand than the previous coups. We have destroyed over 2 million jobs in 2 years, making unemployment rates soar: according to official statistics, Brazil currently has a total of about 14 million people unemployed.[3] In rural areas, makeshift tarp camps have cropped up again by the side of roads and in the city the struggles of the homeless grow each day. These are the manifestations of a people who are fighting for the right to land, work, education, health, and life for centuries.

This book chapter aims to reflect on a) the role of transnational corporations, especially those that are part of so-called agribusiness, b) the struggles to build agroecology, especially those driven by the MST, and c) the role of the MST's agroecology schools in the agroecological transition.

2   Monster Corporations and the Fetishism of the "Green Revolution"

In other texts, we have already commented on the offensive of transnational corporations since the 1960s.[4] In his 2013 book "Mass destruction – geopolitics of hunger", Jean Ziegler calls capitalist corporations "Tiger Sharks". It is a very suggestive name to represent the offensive of corporations in all spheres of our life. As we have seen, Mészáros (2004) calls them "monster corporations". Others prefer to call them "Octopuses", full of tentacles grabbing territories and

---

2   See Ziegler (2013), Mike Davis (2007), Netto (2008) and Lima Filho, Novaes and Macedo (2017).
3   Of course, these state statistics tend to underestimate the rise in unemployment and underemployment. About this, see Mészáros (2011).
4   See Novaes (2017), Novaes et al (2015).

people. Acting as true National States, many of them stronger than most countries, transnational corporations aim their tentacles or their voracious and sharp teeth at everything and everyone.

At the turn of the nineteenth century to the twentieth century, as Lenin (2010) showed us in "Imperialism – the Highest Stage of Capitalism", large capitalist corporations first emerged. We are no longer facing the competitive capitalism recorded by Karl Marx in "Capital", when he described and analysed the capitalism of the first industrial revolution.

In the 20th century, giant monopoly and oligopoly capitalist corporations launched new products on the market, new ways of managing the workforce, created new technologies, promoted wars, overthrew governments, murdered leaders of social movements, stole lands, and caused numerous socio-environmental disasters, among other crimes. The film "The Corporation" shows us the rise of corporations and their tentacles in all dimensions of life. Legally they are impersonal, but behind them there are billionaires, pension funds, and managers who seek the incessant self-valuation of capital.

From the 1960s on, within the project of expansion towards Latin America, China, and the destruction of the Social Welfare State in Europe, capitalist corporations go on a new onslaught. Through technological development that has generated new products and new work processes, with scientific research to manipulate the human mind for consumption and increase the engagement of workers on the factory floor, "tiger sharks" now have a true system of social control, a kind of "Corporate Big Brother" in parallel to the information systems of the National States. Nowadays, the control carried out by the capitalist corporations in our lives is appalling. They control everything we do at work, our email, uncover consumer habits with spy sites, and use drones to kill and deliver. On *Facebook* they follow your "image" and act quickly if any problems come to pass.[5]

One of this chapter's authors has been in several construction works funded by the "Growth Acceleration Plan – PAC". There, one can see the enormous power of large corporations in production and social life in general. We visited the plants of Santo Antônio and Jirau (Rondônia). The contractors exploit workers, confine them to poor housing, steal the lives of riverside dwellers, and promote mass displacement of workers to the construction works. The works lead to numerous impacts on fauna and flora, the emergence of prostitution centres, abandonment of pregnant women and the displacement of people

---

5 For more texts on transnational corporations, see Petersen (2013), Sevá Filho (2013), Campos (2009), Bruno (1999), Bernardo (2002), Sebastião Pinheiro (2005), Fontes (2010), Ploeg (2008), Pinheiro Machado and Pinheiro Machado Filho (2013) and Panitch (2014).

affected by dams, among many other aspects that we cannot develop in this chapter. As arms of capital, NGOs, Institutes, and Foundations emerge in the regions to soften the impacts of the works.[6]

In the name of "progress", of making "a new Amazon", of "employment for the development of the region and the country", of "attracting industries", the tiger sharks aim at the "self-valuation of capital" – capital that is increasingly financialized.[7] With the right hand they earn exorbitant profits and steal whole territories, with the left hand – "returning" one billionth of what they steal from the region, generate "local jobs", heat the economy, build hospitals, day-care centres, and schools, using the discourse of "corporate social responsibility".

Likewise, many capitalist corporations of very high calibre in the agrochemical, transgenic, tractors, and agricultural implements industries and in the commercialization of commodities produce various damages to the working class, such as land theft, poisoning of producers and consumers, increase of allergies, indebtedness of small and medium producers, etc. One of these corporation's websites show them as producers of "agricultural defenders", stating that they contribute to "ending human hunger". In Minas Gerais, Vale, co-owner of Samarco, generated a true corporate social irresponsibility in 2015. Also in 2015, the corporations promoted another "irresponsibility" in Brazil, throwing 2 million workers into unemployment.

According to Bhagavan (1987), who analyses the so-called Green Revolution and its fetishism:

> Irrigation, fertilizers, pesticides, agricultural mechanization, and facilities for grain storage are the essential technological ingredients in the Green Revolution's high-yield-variety (HYV) strategy. Good input prices for farmers are an essential economic ingredient. The uneven availability of these technological and economic ingredients has produced a large increase in disparities between states and between the different classes of farmers

---

6 Michel Torres (2017) shows us how the arm of corporations, that is, Foundations, Institutes and NGOs a) produce naïve reports on how to "solve" social issues; b) hold seminars, forums, meetings, to build and outline a social intervention strategy, c) formulate policies for privatization of social assistance, concessions, creation of charter schools, public-private partnerships, etc.; d) formulate strategies to increase the performance of public servants; f) train or meet managers, secretaries, supervisors of education systems and social assistance managers monthly to talk about educational strategies, "solidarity development" and "local development".

7 To learn more about financialization of the economy, see Delgado (1984), Villaça (1986), Coutrot (2005), Godoi (2006), Fattoreli (2007), Campos (2023), Brunhoff (2009), Chesnais (2010 and 2011), Lapyda (2011), Fix (2011), Arantes (2012) and Galzerano (2016).

in the states. The states with good irrigation, where middle-level farmers make up the majority of farmers had positive annual growth rates in cereal production. Most of the Green Revolution occurred in these three states. States with little irrigation, in which small and marginal peasants and landless workers making up the rural majority, had very low growth rates in cereal production. Generally speaking, no more than half a dozen of India's 22 states (excluding the nine federal territories) have benefited from HYV's strategy, and among them benefits mainly reached middle-sized and large farmers.

BHAGAVAN, 1987, p. 63–64 apud MÉSZÁROS, 2004, p.140[8]

The consequences of the "Green Revolution" were narrated by researchers, documentarians, scientists, and organic intellectuals of social movements, among others.[9] Ziegler (2013), expounds on the "Green Revolution" and shows us the irrationality of commodity production, that is, production focused on capital accumulation and all the consequences that the capitalist mode of production brings to humanity.[10] In dialogue with Ziegler, Rogério Macedo (2015) observes that there is a "destruction of the workforce" and this has a name: humanitarian catastrophe. He states:

> The phenomenon has two dimensions: one systemic and one specific. The first concerns the conversion of the whole system of capital into a machine of mass destruction, by subtracting the minimum conditions of reproduction from the global working class, a process governed by the classical and absolute law of capitalist accumulation. In this dimension, it is called the workers' destructive systemic complex, a situation

---

8   For a rescue of Mészáros' criticism of destructive production, see Mészáros (2002), Mazalla Neto (2013), Rego (2016) and Novaes (2010). For the limits of "sustainable development", see Foladori (2001) and Foster (2005).
9   For the Marxist contribution to socio-environmental issues, see Duarte (1986), Foster (2005), Frederico (2007), Lowy (2003), Kovel and Lowy (2003), Altvater (2007), Sevá Filho (2013) and Sevila Guzmán (2013), Rego (2016). For the Marxist contribution to the debate on agroecology, see Rego (2016), Guhur (2015), Novaes, Mazin and)Santos (2015, Guzman and Molina (2011) and Rego (2016). For the contributions of the left to the debate on agroecology and environmental issues, see Sachs (1986), Primavesi (1986), Petersen (2013) and Ploeg (2008) and Marques (2015).
10 Ziegler is Swiss and was the Special Rapporteur on the Right to Food of the UN, among many other UN jobs. His book has been widely used by all those interested in understanding the role of corporations in the current phase of capitalism and for understanding the increase in world hunger. For the limits and potentials of Ziegler's theory, see the preface to the book made by José Paulo Netto (2013).

aggravated by the determining presence of the structural crisis. The second dimension consists of a specific part of the aforementioned complex (also governed by the general law of accumulation) which is immediately responsible for blocking the positive aspects involved in the growth of food production and marketing. I will call it the hunger and degradation of eating habits complex. Therefore, they are two mutually determined dimensions, one contained within the other: all profoundly destructive, determined by the structural crisis, led to this by the globalization of capital.

Schematically, one could say: the mass destruction of the workforce is the consequence; the aforementioned complex of hunger and the degradation of eating habits is part of the system of capital; its mechanisms are hunger epidemics and the degradation of eating habits.

MACEDO, 2015, p. 311–312

If the "traditional" production of goods were not enough, now transnational agribusiness corporations have a "green" sector, which we could call "green merchandise". Capitalist corporations have realized the gain to be had in new market and adapted to "environmentally sustainable" causes. As everything in society turns into a commodity, the "green" agenda has attracted the middle classes and a portion of the population that is aware of the risks of the Green Revolution package to some extent.[11]

3 Fights for Agroecology and the MST's Agroecological Agenda

The dictatorship that ruled Brazil from 1964–1985 destroyed the social movements of the 1930s-60s. In the second half of the 1970s, workers around the country engaged in various struggles. They took place all over the country: fights against hunger, for housing, employment, better wages, better working conditions for the civil service; bank worker struggles, struggles for land and homes, day-care, basic sanitation, struggles for education and the democratization of public school, struggles of people affected by dams, etc.[12]

---

[11] For the limits of the "responsible consumption" cause, see the book organized by Mazin, Novaes and Santos (2015). It is also worth remembering that the dominant theories on the environmental issue have placed the "blame" for environmental destruction on individuals, without of course placing the spotlight on the main determinant of destructive production: capitalist corporations.

[12] See – for example – Sader (1988) and Dal Ri and Vieitez (2008).

In the end, capital came out victorious with its "gradual, slow, and secure transition". We didn't get the direct elections we asked for in the "Diretas Já" movement.[13] Capital was in control of this transition, to the point that Florestan Fernandes (1986) wondered if we were really entering a "New Republic" phase.[14]

4   Class and Gender Issues in Agroecological Struggles

In March 2017 women from various social movements in the countryside took to the streets against the calamitous social security reform, sexism, capital, and agribusiness. Everything leads us to believe that we are living in times of struggle, that the workers of the countryside and the city will rise up and fight against big capital's new offensive.

In March 2006, after much organizational planning, women in southern Brazil decided to destroy Aracruz's GMO nurseries. They were labelled "anti-progress", "Luddites", "vandals", and "rioters" by the capitalist media (Novaes, 2012). We tried to show that they fought for food sovereignty, for the decommodification of seeds, and unconsciously for a Science & Technology in favour of life and social movements. After that, many actions were carried out against various corporations, against the capitalist state, and against the current division of labour in rural settlements and in the home.

Researchers such as Maria Orlanda Pinassi and Kelli Mafort (2012), Bruna Vasconcellos (2015), Emma Siliprandi (2009), and Marcia Tait (2014) have shown us that in these times of struggle, agroecology is linked to women's economic autonomy, demands for political participation in agrarian reform settlements, fight against transgenics and corporations, and a new division of labour between women and men.[15] The Peasant Women's Movement (MMC), the Movement of Women Rural Workers (MMTR), and the Xique-xique Network, are expressions of these new times of struggle.

The MST, which emerged in 1984, has had a gender working group since 1995, making up a kind of struggle within the struggle. There are many causes within the larger struggle for land, such as environmental and gender struggles, struggles for cooperation against transgenics, for food sovereignty, etc ... Today, all instances of the movement are made up of both men and women.'

---

13   Diretas Já was a popular political movement with the goal of restoring direct elections for President of the Republic of Brazil during the military dictatorship.
14   For this, see also Netto (2010), Sampaio Jr (2013), Minto (2015) and Deo (2014).
15   For the debate on transgenics, see also Benthien (2010) and Moura (2014).

The challenges of female peasants and rural workers are still immense, but they have come a long way. They fought against the dictatorship, but that was forgotten. They fought for the "visibility" of domestic work, for substantive equality between men and women, fought for social rights in rural areas and against gender oppression. In a text about the relationship between Associated Work and Agroecology, Bruna Vasconcellos (2015) says:

> The approximation between Associated Labour and Agroecology, from a feminist perspective, are potential spaces for the transformation of gender relations, not only because they question capitalist forms of production, but also because they represent the possibility of rethinking the reproduction of life, their place of work, knowledge traditionally considered to be female, and the role of women in capitalist society. From the approach to the criticism of the Feminist Economy, Agroecology and Associated Work are seen as potential spaces to re-signify reproductive work. They are the possibility of re-articulating the production and reproduction of life. And yet, it is precisely at the rupture of this division where it is the most difficult to perceive the changes taking place.

Mafort and Pinassi (2012) note that:

> Thus, [according to Mészáros], we live in a historical period in which the socio-metabolic system of capital can only affirm itself by making human beings completely ineffective. And the issue of pesticides and transgenics, as an indispensable increment to the logic of this system, is perfectly inserted in a complex that only cares about increasing productivity and, consequently, its self-reproduction. In this context, the human need for food is absolutely secondary, as well as the fight against hunger and poverty, which is used merely as a pretext for the appreciation of capital involved in agribusiness.
> PINASSI and MAFORT, 2012, p. 82

And they add, bringing the relationship between class and gender from the perspective of Via Campesina:

> the feminist perspective of Via Campesina presents an innovative component to the scenario of class struggle. The actions they carry out against pesticides, transgenics, and agribusiness go far beyond criticism or a mere demand; they are political, aggressive ways of confronting the State and big capital, because they expose the absolute limits of the system, the

most nefarious character of the structural crisis: its total incompatibility with life on a broad spectrum.

Moreover, to the extent that these demonstrations usually suffer all sorts of accusations within their own organizations, the boldness of these working women also exposes the deformed sociability of the class to which they belong, a class that, denying self-criticism, unfortunately begins to negotiate, through the State, with their own tormentor, and that of all humanity. Finally, these women demand that we urgently think of a radical alternative to the system, an alternative that constitutes the realm of freedom and substantive equality.

PINASSI and MAFORT, 2012, p. 88[16]

We saw in Novaes et al (2015) that – if capital promotes formal equality – anti-capital social movements fight for the construction of substantive gender equality to some extent combining with the struggle to overcome the capitalist state and exploitation of labour. It is no coincidence that the women of the MST organize struggles for economic independence, not subordination to their husbands. At the same time, they were involved in class, gender, ethnicity, and environmental issues, in an interesting intertwining mesh that deserves more research.[17]

5   The People's Agrarian Reform and the Construction of the Revolution in Latin America

Agroecology will certainly not advance if it does not conquer land. Without extensive agrarian reform, there is unfortunately no agroecology. Without overcoming labour exploitation and alienation, there is no agroecology. As we saw in the previous pages, without the advance of feminism, there is no agroecology. Following this line, without a complete decommodification of society and without food sovereignty, there is no agroecology. Taking over the means of production is a vital task in the 21st century. More than that, the take-over and control of the means of production by workers with a view to building a

---

16   "When we eat poisoned food and breastfeed our children, instead of feeding life, we transmit death. However, the same government that campaigns to encourage women to breastfeed finances the agribusiness that produces poisoned food for the poor, contaminating the milk of most Brazilian mothers". (MST, 2010, apud Pinassi and Mafort, 2012).
17   See also Angela Davis (2014) and Andreia Galvão (2011).

society governed by totally decommodified, freely associated producers is a vital task in the 21st century.

The MST has advocated for a People's Agrarian Reform. Everything leads us to believe that the Brazilian property-owning classes will not accept an agrarian reform, much less an agrarian reform with a popular character. As a great producer of wealth and misery, Brazil has become one of the largest silos of humanity, but also one of the largest silos of misery. Brazil produces corn for pig and chicken feed, but it does not have corn to feed the children of the working class.[18]

According to David Harvey (2004) and Walter Gonçalves et al (2016), Brazil is one of the central stages of "accumulation by spoliation". Theft of public lands, illegal enclosure of lands, theft of lands from squatters, smallholders, Faxinalenses,[19] etc. have become more common than we think.[20] In this sense, food sovereignty, that is, the fight against the production and export of commodities, gains a primordial role insofar as what is at issue is the adequate feeding of human beings, and not "feeding capital profits".

In the context of this new phase of capitalism, everything points to the emergence of numerous struggles against the closure of schools, struggles for land, for housing, cheap and quality public transportation, access to public universities and public health care. Meagre, hard-fought republican achievements for workers are being destroyed, in a kind of "de-proclamation of the republic".

Faced with this context of a capital offensive and the destruction of everything that is public, what is the challenge for anti-capital social movements? First of all, in the case of Brazil, comes the re-establishment of our democracy and ending this institutional coup as soon as possible.

After that, anti-capital struggles must advance. It may appear that fighting against the closure of schools, against the destruction of public health, fighting for land, housing and better wages/labour rights are reformist struggles. But in the context of capital's offensive manoeuvres, they gain a radical character, however difficult that may seem. But we believe that our struggles will need to

---

18 The marks of our colonial matrix based on the latifundium – immense rural properties, sometimes the size of small states and countries -, production facing exportation, and slave labor are "engraved" in the country to this day. For this debate, see Prado Jr (2002), Sampaio Jr. (2013), Ziegler (2013), Macedo (2015), MST (2014) and Deo (2017).
19 Faxinal is a traditional peasant system of agricultural and animal husbandry production found in Southern Brazil, especially in the state of Paraná. It is characterized by the use of common lands by a Community to produce. fa Faxinalenses are members of these communities.
20 See the interesting article by Walter Porto Gonçalves et al. (2016).

advance towards more precise anti-capital causes: self-management, cooperation, decommodification, ecosocialism, land for work (and not land for business), food sovereignty, substantive equality, and an education beyond capital. Fights for the withering away of the capitalist state and its bureaucracy, the unification of the struggles of the working class and communal property, and the construction of our revolution are good examples of what we are theorizing. Without them we can hardly walk towards a revolution in Latin America.

In the absence of these causes, the proprietary classes may even yield here or there, but the essence of the sociometabolism of capital will be preserved. In the absence of these causes, the struggle for land will become family farming, the struggle for a roof will become at most a segment of the Minha Casa Minha Vida Program,[21] all under the command of the corporations. The struggle for agroecology will remain trapped to the arena of "responsible consumption" and similar ideas.

Our struggles will also not be restricted to electoral contests and campaigns. As Mészáros (2008) warns, the struggles of the 21st century should be based on extra-parliamentary causes:

> the original and potentially alternative extra-parliamentary force of labour has become, in the parliamentary organization, permanently disadvantaged. Although this course of development could be explained by the obvious weaknesses of organized labour at its inception, arguing and thus justifying what happened, in the present circumstances, is only one more argument in favour of the dead end of parliamentary social democracy. For the radical alternative of strengthening the working class to organize and assert itself outside Parliament – as opposed to the defeatist strategy followed over many decades until the complete loss of the rights of the working-class in the name of "gaining strength" – cannot be abandoned so easily, as if a radical alternative were a priori an impossibility.
> MÉSZÁROS, 2008, p. 18

For us, the struggle in the 21st century must be centred on the streets, on working the streets, on critical music, on the *cordeis*,[22] on the unity of social

---

21 The "Minha Casa, Minha Vida" (My House, My Life) Program was a federal housing program created in March 2009 by the Lula administration. The Minha Casa Minha Vida Program subsidizes purchases of houses or apartments for families with an income of up to 1.8 thousand reais and facilitates access to credit and eases hurdles for families with na income of up to 9 thousand reais per month to purchase real estate..

22 Cordel Literature is a folk literature genre, usually written with metre and rhyme. It is usually based on oral performances which are then set to writing and printed in small books.

movements, on popular festivals, etc. always with a view to overcoming alienated labour and its corresponding, equally alienated, form of politics.

History also shows us the need for a revolution. Latin American struggles cannot be solved by the gradualism and reformism typical of the left-wing parties of recent decades. It is not possible to ally oneself to class reconciliation, to ally oneself with capital. We saw the outcome of Lula's government: with the deepening of the economic crisis the workers' aristocracy was "ejected" from the Government, in a perfectly orchestrated parliamentary-legal coup. Of course, a revolution needs a revolutionary theory suited to the 21st century and suited to the specificities of Latin America.

Thus, land in Latin America will only be taken over by peasants, indigenous people, Quilombolas, and rural workers within a revolutionary framework. In its absence, as we have seen, the agroecological agenda of social movements will advance slowly, most likely in the form of a kind of eco-capitalism tolerable by the ruling classes, or in the form of "green" agribusiness.[23]

The advance of agroecology within an ecocommunist and self-management strategy depends on political struggles, or rather the advance of the anti-capital struggles of social movements and the spread of revolutionary consciousness. In Latin America, the revolutionary subject is multiple and more complex than the workers and peasants of the twentieth century. The construction of the unity of the struggles of indigenous people, Quilombolas, peasants, rural workers, the formal and informal urban working class, and the new outsourced working class, will not be easy, but it is essential.

With the degradation of public services in recent years in Latin America, "new characters entered the scene" of urban struggles: teachers in all levels reacted to their precarious conditions; health workers, social assistants, and countless others started their own 21st century struggles and are therefore part of the new working class.

The rise of the indigenous movement in Bolivia, Ecuador and Mexico cannot be neglected either. In Brazil, the process of advancing the new agricultural frontier through agribusiness is leading to the emergence of new struggles led by indigenous peoples, Quilombolas, squatters, etc.[24]

---

23 In Latin America as a whole, numerous struggles for land, housing, water, basic sanitation, health, education, control of natural resources, among others, have broken out. In general, these struggles "stalled" after small demands were met, especially because the offensive of capital did not allow workers to overcome it. On the contrary, it tended to throw workers as a whole into misery or a defensive situation. But it should also be noted that we lack an adequate theory of the transition to communism in the region.

24 According to the CPT Report (2017), from 2010 to 2016 the advance of agribusiness doubled the number of murders in rural areas. We jumped from about 30 to 61. If we include

6 Educational Resistance: The Experiences of the MST Agroecology Centres

The small number of agroecology schools linked to social movements must be understood within the context of capital's offensive.[25] Mônica Molina, Lizete Arelaro, and Wolf (2015) present us with the relentless harassment of monoculture companies – linked to big agribusiness – to rural schools.

In the city of Teodoro Sampaio, a company called "Usina Odebrecht Agroindustrial", using several strategies involving the municipal government, community members, school leaders and agents, has inserted itself into rural schools in the region through the "Social Energy for Local Sustainability Program", disseminating and promoting counter-values among teachers, students and the community. They extoll the "benefits" of agribusiness for the territory, making it difficult to understand the immense contradictions that are hidden in plain sight in agricultural model. One of the most perverse has been trying to convince the youth of the agrarian reform settlements to give up the greatest victory achieved with the struggle for land, that is, allowing for this means of production to proliferate, convincing the youth to sell their workforce to these monoculture companies. They are even able to convince many families to lease their lots to these same companies (Molina; Arelaro; Wolf, 2015).

On the other side of the class divide, the construction of the MST Agroecology Centres in Paraná is linked to the founding objectives of the Movement: "fight for land, fight for agrarian reform, and fight for social changes in the country" and the dispute for the productive agricultural matrix.

In view of an alternative proposal for the education of the working class, the main objectives of the MST Agroecology Centres in the State of Paraná are:

- To be a training space for working-class organizations;
- To be a space where meetings of the MST and other organizations which seek the same goals of social transformation can take place;

---

assassination attempts, the numbers are staggering. Not to mention slave-like labor in the 21st century.

25 We could go further, because the offensive of capital prevents the emergence of social movement schools and at the same time closes schools. See, for example, the excellent documentary "Granito de Arena" on the closure of rural technical schools in Mexico, and the numerous articles that have emerged on the occupations of schools in Brazil in recent years. It is also worth consulting the writings of Section 22, the teachers of southern Mexico, and the teachers of Neuquén (Argentina).

- To be a reference in the development of experiments in the field of agroecological production, presenting concrete results for farmers;
- To be a space for the development of socialist humanist values, matured through collective life;
- To improve methods of technical and political training and schooling since elementary school, as well as in high school and higher education;
- To enable the development of scientific and technological experiments, focused on the peasants' reality;
- To be a space where popular culture is encouraged and experienced, especially peasant culture;
- To be a space where people can live, educate, work, have fun, and build perspectives of the future.

MST-PR, 2004; LIMA, 2011, p. 87

For us, the creation of the MST Agroecology Centres in Paraná represents an important, albeit still under construction, space for: a) the formation of a militant framework; b) the socialization of historical and scientific knowledge produced by humanity; and c) the approximation of rural and urban workers, supporting the construction of collective actions of common interest (Lima, et al. 2012, p. 194; Pires, 2016).

The methodological theoretical foundations that guide the Political-Pedagogical Project (PPP) of the courses developed at the MST Agroecology Centres are based on the political and educational *praxis* of the principles of socialist pedagogy, popular education, dialectical historical materialism, and the *Pedagogy of the Landless Movement* (Caldart, 2004, 2015; Guhur, 2010; Lima et al., 2012; Pires and Novaes, 2016).

Referring to the work of Caldart (2004, p. 315), the landless worker's education has the MST as its main pedagogical source, "as a collective in movement, which is educational and which intentionally acts in the process of formation of the people who constitute it". Within this, the *Pedagogy of the Movement* develops its educational matrix in five dimensions: a) the pedagogy of social struggles, b) the pedagogy of collective organization, c) the pedagogy of the land, d) the pedagogy of culture, and e) the pedagogy of history.

Seeking to articulate work, education, school, and community, the educational proposal of the agroecology courses developed at the Centres also refers to the concept of "socially necessary work" developed by the socialist pedagogy of Viktor Shulgin (2013). In this frame of refence, "socially necessary work" proposes the basis of school life, not as a mere adaptation, training of hands and/or teaching method, more organically and closely linked with teaching.

Becoming increasingly complex, it must be the light that surpasses the limits of the immediate situation, enabling the knowledge of life and the most diverse forms of production.

Following along the lines of influence of socialist pedagogy, we come to Pistrak (2010), for whom complex teaching is not reduced to a simple method that can provide a better way of assimilating content, it is something deeper, which is related to the essence of the pedagogical problem and with the knowledge of real phenomena and their relationships, that is, the Marxist conception of pedagogy.

In this context, the Political Pedagogical Projects (PPP) of the Agroecology Centres were based on the *Pedagogy of the MST*, as well as the principles and concepts developed by Soviet pedagogues, including Pistrak and Shulgin. In this perspective, work, self-organization, and the relationship with the community are principles behind its Political Pedagogical Project (PPP) and its Methodological Project (ProMet), as we can see in the case of the José Gomes da Silva School (EJGS) presented in Table 6.1.

Through these principles, it is proposed that education be developed from a kind of pedagogical work that develops experiences, decision-making, work, and learning in a collective and participatory dimension, which has links with the working class, which is critical, and seeks to further the MST's organicity and demands.[26]

Starting from organicity, students who participate in agroecology technical courses, for example, will be organized in Base Centres and work teams. Work is seen "a trigger for new learning, through the practice-theory-practice paradigm, producing knowledge about reality" (PPP, 2010, p. 11). For a more didactic understanding of the role of work teams in the educational process and in the structure of the course and the Centre, Table 6.2 below presents the teams created for a class in a technical high school course in agroecology held at the José Gomes da Silva School.

The students in each team propose how they will insert themselves in the local reality and in the course itself. The first way to do this is through

---

26   The term organicity is widely used in the internal debates of the MST. Its meaning and content include: expanding participation, building awareness with participating families, training militants – cadres, having political control of geographical space, implementing organic circles, keeping permanently vigilant, removing enemies, accumulating forces. All this will help in developing a strategy in the political struggle for agrarian reform, creating the conditions to wage a political dispute in Brazilian society. For more information on the organicity of the MST read: Working Method and Popular Organization. National Training Sector – MST (2005).

TABLE 6.1   Pedagogical principles of the José Gomes da Silva school

Values	Description
Collective direction	All instances will be made up of commissions of workers with equal rights and power. As a matter of priority, decisions will be taken by political consensus.
Division of tasks	Stimulating and applying the division of tasks and roles among the subjects of the collectives, valuing the participation of all and avoiding centralization and personalism.
Professionalism	All members of the sectors and collectives must fulfil their roles with professionalism. Professionalism should be considered in two aspects: a) to take on the struggle for land and the organization of the Movement as a militant profession. To love and dedicate oneself in body and soul to it; b) To be an expert, seeking to improve oneself continuously in those roles and tasks you are assigned to, in view of the Movement's organic whole.
Discipline	Apply the principle that discipline is respect for the decisions of the collective, including the fulfilment of schedules, but above all tasks and missions.
Planning	Apply the principle that nothing happens by chance, but everything must be evaluated, defined, and planned from the reality and objective conditions of the organization.
Study	To encourage and dedicate oneself to studies of all aspects that concern the activities of the Movement. Organizations that do not form their own political cadres will not have the autonomy to conduct their struggles.
Connection to the masses	The permanent connection with the masses of workers is what guarantees the advancement of struggles and the application of a correct political line. We must learn the aspirations and desires of the masses and, based on their experience, correct our proposals and decisions.
Criticism and self-criticism	Always apply the principle of critical evaluation to our actions and, above all, have the humility to undertake self-criticism, seeking to correct errors and find solutions.

TABLE 6.2   Role of work teams in José Gomes da Silva School

Teams	Description
Health/Sports/Leisure	This group will have the task of organizing health-related activities and preparing natural remedies; in urgent cases of referrals to the doctor (hospital or health post) this team will accompany the patient(s). They will also plan activities that contribute to the improvement of hygiene and cleanliness as part of preventive health, as well as holding seminars on health-related topics. Furthermore, the group is responsible for organizing the cleaning shifts and monitoring collective spaces, ensuring cleanliness, organization, and aesthetics. They also have to coordinate the use of cleaning materials and products. Organizing sports and leisure time with recreational activities for the well-being of the group is also part of this team's attributions. The team should plan diverse activities involving all students to engage physical exercises for the sake of everyone's physical and mental health.
Human Relations	This team has the responsibility to guide and ensure conscious discipline among all members. In cases of indiscipline, educational activities should be carried out in order to raise awareness about personal limits before the collective, thus seeking to overcome them. It will also have the task of ensuring compliance with collective agreements regarding schedules, Course and School rules, as well as good conduct and relationship between all militants.
Communication/Culture and Mystique	This team is responsible for: carrying out energizing activities with the class, especially during class time as well as monitoring and developing cultural activities in the allotted time. They will also be responsible for preparing the "news time" and organizing the information board, as well as the ornamentation of educational spaces. The group will also be responsible for the use of electronic sound and video equipment with the unit coordinator by EJGS-ITEPA.

TABLE 6.2 Role of work teams in José Gomes da Silva School (*cont.*)

Teams	Description
Reporting and Systematization	They are responsible for setting down the memory of the course, creating and systematizing daily reports on the development of daily activities. The group also records advances made and creates challenges to be overcome by the class in organizational, learning, participation and practical aspects.
Production and Infrastructure	This group assists in the planning and monitoring of work time, as well as monitoring the school's physical structures and arranging for someone to repair them when necessary. The team also has the responsibility to plan the school's garden.

SOURCE: AUTHOR'S ORGANIZATION, BASED ON PROMET (2010)

self-service, through which students are fundamental in processes of maintenance, production and care for the school's people, structures, and equipment. They are also essential to the educational processes of school time, where they are responsible for discipline, commitment, and respect of the class towards educators and in the various educational times.

Second, by participating and contributing to the school's productive units, their insertion in the productive units is monitored of bythe person responsible for each sector and the Political Project Council (PPC). The objective of participation in these activities is to enable students to have practical knowledge, which must be critically analysed and improved, in addition to contributing to productivity and, consequently, to the self-sustainment of the school and the course.

Third, the organization of educational times in line with the other spheres of teaching and learning in the work teams (self-service) and in the school's productive units (self-support), as shown in Table 6.3.

Taking Shulgin's (2013) notes on "socially necessary work" as a reference, it is observed that the PPP of the MST Centres proposes three important basic points. The schools are: 1) oriented towards economic and life improvement; 2) pedagogically valuable, and 3) in accordance with the strengths and particularities of the adolescent students.

TABLE 6.3   Description of the educational times of the "earth revolutionaries" class

Educational time	Description
Class time	A time when the disciplines and thematic axes are developed in the curriculum's various areas of knowledge. The thematic axes refer to different disciplines according to the students' grade and technical topics, among others. There may be some changes, as it is necessary to reconcile this to the educators' schedules.
Reading time	Activity aimed at reading and individually directed studies, guided by the need of each student to appropriate certain subjects, in order to build an appropriate method for the study and development of reading habits, research and intellectual development, as well as providing moments of socialization for the class as a whole.
Working time	Working time is defined in view of the internal demands of the EJGS, contributing to the production and maintenance in the various sectors/units of the centre/school and activities necessary for the well-being of the community and the formation of social and humanist values. In this sense, working time must happen as a formative element that develops collectivities, organization, and cooperation. The insertion of students into working time also fulfils the role of conducting productive research which may contribute to the planning of activities and the organic construction of each sector.
Workshop and seminar time	Intended for the learning and development of specific skills related to the class's training focus. Students master new activities during this time. It can also be used to qualify the work done in the production units. It is organized according to the dynamics of the classes and readings.

TABLE 6.3 Description of the educational times of the "earth revolutionaries" class (*cont.*)

Educational time	Description
Mystique time	Mystique is the soul of the Landless identity. The EJGS has the task of rescuing the love of work and the belonging of the student and the Landless community to the working class. Mystique is more than a time, it's an energy that permeates everyday life. That's why you need it at the beginning of large activities and you bring it up various times during the day. This activity is the responsibility of the base nuclei. One must learn to work and experience the mystique, cultivate the struggle of the workers, and honour important dates and achievements. It is also the time when the base and information nuclei have their meetings.
Written reflection time	Aimed at recording the experiences that each student extracts from the educational process at the school and the course, which will contribute to their militancy. It is the moment in which students reflect on their daily practice and the challenges to be overcome. To do so, each student will have a specific notebook, in which they will record their activities daily, according to the organization of each subject. The same will be requested by the pedagogical coordination for weekly monitoring.
Culture and leisure time	Intended for cultural activities, plays, dances, trips, music, peasant culture events, and others. The communication and culture team have the responsibility of coordinating this time. This time is organized according to demands presented by the class.
Base nucleus time	Aimed at the discussion and general referrals of the class and the course. It is also a space for study and debate on behalf of the self-organization of students in the organic processes of the EJGS and the MST.

TABLE 6.3 Description of the educational times of the "earth revolutionaries" class (*cont.*)

Educational time	Description
News time	A moment designated to follow the news through television, newspapers, and magazines, making critical reflections on the facts as they are are reported by the media. Videos, documentaries and lectures are also included. These activities are the responsibility of the culture unit and the communication team with CPP guidance.
Complementary study time	The aim of this moment is to provide students with space for self-organization, individual and/or collective studies, coursework, and other activities.
Joint project time	This time is dedicated to care for the school, with the appreciation of small, everyday tasks, and beautification of the collective public space. It is also used to carry out general cleaning of the school premises. It is carried out according to the dynamics and demands of the EJGS as a whole.
Community time	The objectives of this time are: To carry out activities delegated by the organizations the students are a part of; to commit to the execution of alternative production lines; to develop activities guided by the educators and the pedagogical coordination; and to develop field practices. This work is constantly evaluated and redeveloped. The students will carry out activities that will be accompanied by the political pedagogical coordination of the course, technicians, MST sector collectives, and brigade directors.

SOURCE: AUTHOR'S ORGANIZATION, BASED ON PROMET (2010)

The educational times, as described in the table above, reinforce the principles that "school is a place of human formation, and therefore the various dimensions of life must take place in it, being worked on pedagogically". Thus, "educational time contributes to the organization process of students, leading them to manage interests, establish priorities, and take responsibility" (PPP, 2007, p 12).

Beyond its role as a structural part of the future technician's formation, educational times have the characteristic of being holistic, letting students experience and understand the school and the course as a whole through the practical principle of "socially necessary work". Therefore, the agroecology technician courses offered by MST Centres have the objective of:

> training professionals committed to the implementation of sustainable rural development models in their multidimensional form, that is, professionals who have an understanding of a variety of dimensions of knowledge such as organic agriculture, biodynamics, and permaculture, among others.
>
> PIRES, 2016, p.115

The attention given to "develop the habit of reading, research, study and written preparation", in order to "promote the integration between the different levels of knowledge" is also noteworthy. In the same vein, it intentionally trains "conscious and socially committed" professional researchers with "a humanistic view, ethical and holistic values, besides being active subjects in the struggles of social movements" (Guhur, 2010; Lima, 2011; Pires, 2016).

Fostering the interrelationship between work, self-organization and community relations, the courses work in a regime of alternation, articulated in two complementary stages: school time (TE) and community time (TC), which to a certain extent can be understood as an intentional organicity with respect to overcoming the forms of teaching that Shulgin (2013) called "sitting complexes".[27]

In this sense, Guhur (2010) highlights the following about the MST's courses:

> The MST's formal courses are organized in a regime or system of alternation, combining periods of activities in the school (and also field activities promoted by the school), School Time (TE), which is a face-to-face time/space; and periods in the student's home communities, Community Time (TC), which can be understood as a semi-presential time/space. It is important to note that "home community" here is directly linked to the social movement to which the student belongs; it is during TC that

---

27 Sitting complexes are the kind of teaching promoted by educational institutions based only on theoretical teaching and textbooks. They refer to a reading of reality, but they are not a part of practical experience of the students' reality (Shulgin, 2013).

> the Pedagogy of the Movement, (...), acts most strongly. Thus, "for the Landless, the MST is the teacher during TC".
>
> ITERRA apud *GUHUR*, 2010, p. 156

School time activities are characterized as the organic and collaborative participation between the Pedagogical Political Coordination, the families residing in the Centre, and the students themselves in conducting the pedagogical processes of maintenance, production, and self-organization of the school and teaching.[28]

Dominique Guhur (2010, p. 156), coordinator of the Milton Santos School, says:

> During TC, students develop works directed by the school, such as: readings, recordings, field research, internships, experiments, and complementary courses. Furthermore, they must actively participate in the organicity and struggles of the Social Movement they are a part of, and maintain the roots in the community or collective of origin, participating in their activities (sometimes, the responsible Social Movement can send the students to another community during a given TC, or the students can remain in school, contributing to its construction or maintenance).

It is understood that TC is a time when students, following the guidelines learned during educational time, are inserted in their home area with the intention of bringing together the acquired knowledge, setting up the conflict created by the contradiction of the real with the ideal, that is, the transition from the "Green Revolution" to the agroecological paradigm.

The joint formative process that takes place in both TE and TC highlights the importance of experienced and systematized educational spaces as an opportunity for the working class to take possession of the knowledge that was taken from it, but also of locally-created knowledge, born from the perspective of those living amidst the contradictions of capitalism.

> Generally, formal professional education courses – taken here in a broad sense – represent the *locus* "where the MST, as a whole, expresses its conception of what a school is, in its tensions, contradictions, and in the

---

28 For a more focused reading on the issue of participatory management of the MST Agroecology Centers/Schools in Paraná, see the dissertation by Laís dos Santos (2015).

reaffirmation of principles, generally in contrast to the logic of its partner institutions".

MST *apud* LIMA et. al. 2012, p. 193–194

Thus, using the *Pedagogy of the Landless Movement*, and the principles of socialist pedagogy and dialectical historical materialism, the Agroecology Centres are islands of resistance surrounded by huge green deserts. They aim to train technicians, researchers and militants to face the green desert and bring another development matrix for agriculture to fruition, based on a technoscientific foundation called Agroecology.

CHAPTER 7

The Political Economy of the "Green Revolution", Agroecology and the MST Agroecology Schools

1 Introduction

This chapter aims to reflect on the political economy of the "Green Revolution", the debate on agroecology, and the MST's agroecology schools. To achieve this objective, in the first section we present a critical analysis of the so-called *Green Revolution* fetish, characterized by the concentration of land in the hands of few owners and transnational corporations, control of production and distribution of seeds, tractors, synthetic fertilizers and pesticides by corporations, besides the many socio-environmental problems it caused: cancer, destruction of the immune system, increased unemployment, indebtedness of small farmers, etc.

In the second section, we present the MST's concept of agroecology, a concept that goes beyond the traditional meanings of environmental preservation and/or organic production. Agroecology is a political-economic and cultural concept that mobilizes the MST, in the sense of forging new social relations that include: associated work; the appropriate use of agroecosystems; the reconstruction of agriculture through popular agrarian reform, with democratic and participatory management in cooperative and agroecological production systems; gender issues; debates on decommodification; and agroecology education. The entry of agroecology into the MST agenda led to the creation of several technical schools, which I will briefly present later in the chapter.

2 Capital's Agriculture Campaign and Destructive Production: The Political Economy of the "Green Revolution"

Since the 1960s, capital has been carrying out an offensive to transform agriculture. Its ideologues call it the *Green Revolution*. Novaes (2012) calls the same phenomenon the *political economy of the Green Con*. The political economy of the green con is characterized by a) a new cycle of primitive accumulation (land grabbing and theft, murder of leaders, theft of indigenous knowledge); b) the growing concentration of land in the hands of transnational corporations;[1] c) mergers and acquisitions in the seeds and pesticides sector, with a few large corporations from northern countries exercising near-complete domination of the production and distribution of seeds and pesticides; d) a biotechnological "revolution" (new pesticides and transgenic seeds); e) the absence of autonomy for small producers, increasingly working for banks and linked to agro-industrial corporations.

According to Costa Neto (1999), over the last 40 years a drastic restructuring of the input production and industrial transformation sectors took place alongside transformations in financing and credit institutions and mechanisms, commerce circuits, and market structures. Educational, research, and technical assistance institutions were progressively adapted to follow this model, with a view to training agronomists, researchers, specialists, extension workers, and other professionals within the philosophy of the *Green Revolution* (Serafim, 2012).

From a historical perspective, Pinheiro Machado (2009, p. 1) observes that:

> since Liebig, 1848, the capitalist industry has seen in agriculture an excellent source for the reproduction of capital, and, from that point on, agronomy schools around the world do nothing other than teach students to apply synthetic fertilizers, pesticides, and use heavy machinery.
> PINHEIRO MACHADO, 2009

Gonçalves (2008, p. 20) defines the Green Revolution as a:

> technological package that resulted in the industrialization of Brazilian agriculture and, consequently, in an increase in agricultural production, an increase in the exploitation of social surplus value, and in concentration of capital in its various parts, especially commercial, agro-industrial, industrial, financial and real estate. The organization and diffusion of the 'Green Revolution' was bountifully 'watered' by significant sums of resources from public, private, and multilateral development agencies. Its implementation attacked and continues to attack so-called 'traditional' economic and organizational logics, including those of peasants, indigenous people, Quilombolas, Faxinalense communities, etc. Thus, elements such as agricultural mechanization (tractors, harvesters,

---

[1] The political economy of agribusiness has significantly altered the land market and the landscape of Brazil through the purchase or lease of land by financialized foreign corporations (Oliveira, 2010).

> processing machines, plows, harrows, irrigation pumps, sprayers, agricultural aircraft), petrochemical inputs (fertilizers, insecticides, herbicides, maturers, antibiotics, micro-nutrients, plastics for agricultural use in irrigation and crop protection), 'improved' plants and seeds (hybrid, re-engineered and transgenic), as well as the rise of agro-industrial companies, have become structural elements in a changing agrarian space. [...] In addition, [...] it subjugates the work of rural producers to the dictates of agribusiness companies, increasingly organized and taking part in an oligopolized world agricultural trade dominated by 'tradings companies'.
> GONÇALVES, 2008, p.20

The logic of productivism, that is, maximum profit in the shortest possible amount of time, has always been present in capitalism but has been exacerbated over the last 40 years. Guided by the tripod of transgenic seeds, agrochemicals/synthetic fertilizers, and heavy machinery, agro-industrial corporations have consolidated a power and domination structure in rural areas.

Shiva (2001) describes this new phase of commodification of capitalism with the terms *new colonialism*, *biopiracy*, and *biocolonialism*. To the author, "while biodiversity and indigenous knowledge systems satisfy the needs of millions of people, new patent and intellectual property rights systems" threaten to appropriate "the vital resources and processes of knowledge of peripheral countries and convert them into a monopoly to the advantage of Northern companies. Patents are therefore at the centre of *new colonialism*. Furthermore, rising productivity in agribusinesses gave rise to diseases such as mad cow disease, avian flu, and swine flu, among others".

It is true that the "Green Revolution" offensive is still hegemonic, but it is also true that its advance did not take place without resistance. In Argentina, at the end of 2013, a fight broke out in the small town of Malvinas, in the province of Córdoba. Monsanto – the American multinational agricultural and biotechnology company and world leader in the production of genetically modified seeds – intends to install one of the largest transgenic seed production plants in the world in Malvinas. Social movements, NGOs, and environmental groups are trying to block the construction of this plant.

The political economy of the Green Con has plenty of consequences: land concentration; increased rural unemployment; soil degradation; compromised quality and quantity of hybrid resources; devastation of forests and other native biomes; impoverished genetic diversity of cultivars, plants, and animals; contamination of water and food consumed by the population; increased allergies, deaths, and disabilities; and increased commodification and proletarianization of the field, in addition to surges in small farmer debt and rural school closures, among others.

Bayer, Basf, Syngenta, Monsanto, Dow are some of the large corporations that control the market for transgenic seeds and pesticides in Brazil. According to Folgado (2013, p. 1), Brazil has been the largest consumer of pesticides in the world since 2008. "The quantities used are equivalent to about 5.2 litres of poison per inhabitant per year. However, Brazil represents only 5% of the agricultural area among the 20 countries with the largest food production". This means that the country's productivity does not justify its leadership position in the poison use ranking. According to data from the Brazilian Institute of Geography and Statistics (IBGE) and the National Health Surveillance Agency (Anvisa), the inappropriate or excessive use of this type of input can cause serious damage to human health and the environment.

The specificity of the Latin American case lies in the fact that the restructuring of agriculture was implemented during a counter-revolutionary process started by the continent's military dictatorships, which reaffirmed the power of the ruling classes, especially that of the landowning forces of the colonial past in association with foreign capital and sectors among the military and industrialists (Sampaio Jr, 2013; Novaes, 2012). In other words, large landowners, alongside the military, parts of the industrial bourgeoisie, the middle classes, the Catholic Church and the US's CIA, led coups that stopped the advance of the workers' struggles in the region.

More generally, the first symptoms of the capital accumulation crisis appeared in the 1970s. In response to this crisis, capital promoted an offensive that could be summarized as follows: a) pressure for the free movement of financial capital; b) productive restructuring of the countryside and the city; c) technological innovations that intensified the production and diversification of goods; d) expansion towards sectors and fields not yet subject to full commodification, such as health and education, with a wave of privatization, which took the cycle of commodification to new heights; e) implementation of processes, including changes in location and outsourcing, that practically dismantled the power of the combative unions of the Taylorist-Fordist eras of capitalism; f) theft of lands from small producers, indigenous people, Quilombolas, etc.

However, the dynamics of capitalism from the 1970s onwards could be called destructive overproduction, linked to maximum expansion and the corresponding profit, engendering an alienating and ever-expanding self-reproduction (Mészáros, 2004). That is, there are profound changes, both in the objective (economic) plane of production, as well as in the subjective reproduction of class relations.[2] In this sense, we can see that any of capital's "civilizing" facets, so vigorously promoted by liberals, have been exhausted with capital's own destructive advance. Regarding social movements and political transformations, a crisis of traditional expressions and class representations took place. This culminated in the emergence of the so-called "new social movements", which no longer aim to criticize the destructive logic of capital and the articulation of its specific flags to the overall class struggle, but the demand to expand the statutes of "citizenship".

According to José Paulo Netto (2008), the banners of the "new social movements" are restricted to the struggle for insertion in the bourgeois state, and not overcoming it as an instrument of class oppression. He states:

> Imperialist corporations implement the erosion of state regulations. "Deregulation" is presented as a type "modernization" that values "civil society", releasing it from the tutelage of the "protective State" – and there is a place in this ideological construct for the defence of "freedom", "citizenship", and "democracy".
>
> NETTO, 2008 p. 17

Especially since the 1980s, with the destructive advance of capital and, consequently, the commodification of all spheres of life, several conflicts began to emerge in Latin America around land, water, seeds, electricity, oil, gas, education, and labour, among others. In southern Brazil, for example, the destruction of Aracruz's Eucalyptus nurseries and the struggles of the Via Campesina against Syngenta Seeds (a Swiss multinational), can be interpreted as confrontation tactics employed by social movements against a new offensive by capital. These struggles occurred mainly in reaction to the destruction of Creole seeds, to the role that science and technology play in the sociometabolism of capital and due to the control that multinationals exercise in all dimensions of workers' lives. Because of its contradictions, the political economy of the Green Con, which restructures the field and industrializes it, furthering the accumulation of capital to a new level, it gives rise to several social movements that try to point out ways to move forward antagonistically to the paths presented by agribusiness.

---

2   For Mészáros: "capitalism, as a mode of social reproduction, is characterized by the irreconcilable contradiction between production for use (corresponding to need) and production for exchange, which at a certain stage of development becomes an end in itself, subordinating all considerations of human use to its completely perverse logic of alienating self-reproduction". (Mészáros, 2004, p. 297).

3   Agroecology in the MST: Beyond the Green Agenda

In common sense, agroecology is associated with caring for nature, ecology, environmental protection, and small-scale vegetable production. However, this way of understanding social reality tends to be in favour of maintaining relations of domination (Mészáros, 2004). For us, the agroecology theorized by the MST is not reduced to an environmental agenda. Caporal and Costabeber (2002) note that no product will be truly ecological if it is produced at the expense of the exploitation of the workforce, even if it does not use certain inputs, such as pesticides, or if it is using new forms of soil depletion or degradation of natural resources.

For Gonçalves (2008), what mobilizes the MST is the denial of the existing agricultural development model in the country, highlighting the need for the preservation and reconstruction of peasant agriculture through agrarian reform, in addition to proposing forms of management and participation of the peasantry in cooperative and agroecological production systems.

Guhur and Toná (2013), members of the Milton Santos MST School of Agroecology (Maringá – PR), note that a more recent and expanded concept of agroecology is in gestation. This school of thought does not see agroecology as a merely technological solution to the structural and contextual crises of the economic and agricultural model. Agroecology is understood as part of the strategy of fighting and confronting agribusiness, the exploitation of workers, and the depredation of nature. This way of thinking agroecology includes caring for and defending life, food production, and political and organizational awareness, as observed by the Via Campesina and the MST (Toná; Guhur 2012, p. 66).

The authors also state that agroecology is seen as inseparable from the struggle for food and energy sovereignty, for the defence and recovery of territories, for agrarian and urban reform, and for cooperation and alliance between the peoples of the countryside and the city. Thus, agroecology is part of the construction of a society of freely associated producers working to support life, a society in which the final objective ceases to be profit and becomes human emancipation (Mazalla Neto, 2013).

The MST's incorporation of the agroecological agenda took place gradually. According to Gonçalves (2008), this renewal of the movement's agricultural technological matrix started catching on starting in the year 2000, when the Movement began to disseminate the idea that squatting and settled families should, first, guarantee the family's food security with a high quality and diversity of products that are, above all, free of agrochemicals and transgenics.[3]

Gonçalves (2008) also states that for the MST, the change in the social, ecological and, above all, technical rationality of families would help to overcome the new dynamics of capitalism in the countryside, based on extremely severe relations of domination, including the presence of transgenic seeds and the schemes of different branches of transnational agro-commercial capital (chemical, food and financial), which put a stranglehold on farmers.

For the MST, agroecology means stimulating the practice

> of an agriculture with no use of external inputs and agrochemical pesticides. Over the years, we should adjust towards this way of producing, avoiding spending money on fertilizers, poisons, and machine-hours. Rather, we should seek to use locally available labour more and in better ways and developing techniques adapted to our reality, so we don't poison ourselves and nature. We should be open to the creativity of our comrades, producing a new technological matrix.
>
> MST 2000, p. 50–51

In August 2013, the permission to use GMOs in Brazil turned 10 years old. In the 2000s, numerous campaigns were carried out against the use of GMOs and pesticides in the country, which included actions such as the occupation and destruction of nurseries, accusations in the media, production of films and documentaries, occupation of corporation headquarters, dissemination of reports by scientists against GMOs, synthetic fertilizers, heavy machinery and pesticides, and campaigns among social movements, researchers from public universities, sectors of Embrapa, and NGOs. It is worth noting that the MST conducted or participated in numerous actions described above.

In order to combat the restructuring of the field and the Green Revolution and disseminate the theory and practice of agroecology, the MST has so far held thirteen Agroecology Days, created numerous agroecology technical courses, integrated high school courses, specialization courses, undergraduate courses and even a master's degree. It is also spreading the gospel of agroecology in elementary and high schools.

The MST's concept of agroecology is complex and differentiated from an analysis that reduces the term to mere environmental protection, as it

---

[3] Seeds that in Brazil were grown illegally and procured in Paraguay and Argentina.

introduces or covers several related elements. Carter and Carvalho (2004) mention cooperation, gender, education, youth, and the environment.

We can raise the hypothesis that the struggle for agroecology is coupled with what Mészáros (2002) calls substantive equality. If capital promotes formal equality, anti-capital social movements are fighting to construct substantive gender, ethnicity and generational equality and, above all, to overcome class exploitation. It is no coincidence that the women of the MST organize struggles for economic independence, not subordination to their husbands. At the same time, class, gender and environmental issues were involved in an interesting and intertwining way (Pinassi; Mafort, 2012, Tait, 2013). Despite being part of the Movement, which provides a practice of struggle and political awareness, MST members are still part of general society and, as such, suffer the influences of the hegemonic ideology.

In the specific case discussed here, they also suffer from the influence and pressure exerted by large transnational corporations, which dominate agro-industrial commerce and production. Thus, not all MST settlements are based on agroecology. By direct and indirect mechanisms, such as propaganda, which creates the fetish of the Green "Revolution", the influence of technicians and agronomists, or by means of more subtle instruments, such as linking credit to the acquisition of the Green Revolution package, capital creates the general conditions for structured production along the lines of the Green "Revolution". This causes the *usual* form of agricultural production, based on the use of pesticides, synthetic fertilizers, and heavy tractors to be used in a part of the movement's settlements. However, one can see that even in these settlements there is already an effort to progressively move to other forms of production. However, inserted as they are in the capitalist market and under capital's current offensive, it is not easy to oppose the hegemonic model.

The State provides credit with low interest rates, subsidies, seeds, privileged technical assistance, and the development of applied scientific and technological research to agribusiness, among other mechanisms and incentives that maintain the Green Revolution standards (MOURA, 2014). At the same time, large corporations and the state disfigure and hinder the general conditions for the collective and associated production of healthy foods.

Several studies highlight the difficulties of technical assistance for family farms, which deal with precarious professionals, few technicians per settlement, poor working conditions for technicians, damaged equipment, lack of specific training in agroecology and stimulation of marketing networks, etc. This situation portrays one face of a State that is minimum for workers and maximum for financial capital. In Marx's terms (1996), the state and governments create the general conditions for the production and reproduction of capital and annihilate, subordinate, or hinder alternative forms of production that can give rise to a new mode of production.[4]

From the point of view of consumption, despite the existence of agroecological marketing networks, one of the challenges for agroecology is the fact that its products are accessible, mainly, to the middle and upper classes of Brazilian society. According to information provided by the head of the seedling nursery at the Milton Santos School, the costs of producing organic vegetables are not much higher than those cultivated with pesticides. The technician may be right when it comes to small plots. We also need to remember that the final price of food grown with pesticides, transgenics and synthetic fertilizers represents only the tip of an *iceberg*. If we consider all the costs generated by agribusiness to the Public Health System (SUS), to consumers who have to buy medicine in the pharmacy because they have been poisoned, the subsidies given to agribusiness, the final global price of agribusiness products is very expensive, even if on the shelf they appear as "cheaper".

Whatever the case may be, the accusations levelled and pressure exerted by the social movements, combined with greater ecological awareness, ended up resulting in public policies aimed at the purchase of organic food for schools, day-care centres, hospitals, sanatoriums, etc. This is the case of the Food Acquisition Program (PAA) and the National School Feeding Program (PNAE). However, even in the municipalities where this policy is applied, the purchase volume does not reach 10% of the total.

Although limited and prone to becoming islands in the capitalist market, the experiences of spreading agroecological and fair-trade practices must be recognized and disseminated by social movements, such as MST's aforementioned Bionatur, based in southern Brazil.[5]

---

4   Sampaio Jr. (2013) makes a good overview of the actions of military governments to strengthen agribusiness, going through the period of "democratization", and even reaching the Lula years in the section "institutionalization of the counterrevolution and liberal adjustment".

5   It would also be important to highlight the arrangement that has allowed the genetic improvement of native seeds in Paraná and Santa Catarina. Torres-Rego (2015) and Luzzi (2007) synthesized the work of greater repercussion of AS-PTA (Family Agriculture and Agroecology) of the Contestado and Centro Sul do Paraná: rescue and multiplication of varieties of native seeds of corn, beans, cassava, rice and potato; implementation of a creole germplasm bank in 2003 and a genetic improvement laboratory in 2004 at the State University of Londrina (UEL), subsidizing experiments to evaluate creole cultivars by farmers together with researchers from UEL; holding creole seed fairs; experiments in ecological soil management in partnership with the Paraná Agronomic Institute (IAPAR) and the Brazilian Agricultural Research Company – Agrobiology; organizing agroecological product markets; commercialization of ecological yerba mate with the brand Sombra dos Pinheirais by an association of farmers; creation of a consortium of five community organizations for

4   The Creation of the Agroecology Schools

Agroecology reveals a new way of producing, as previously explained. Thus, the MST decided to create schools, alternatives to state schools, that would train technicians, who are not solely technical but also political specialists, according to their needs. To create the schools, the Movement relied on its educational proposal and years of experience, but it was also necessary to count on the involvement of its educators to develop a new curriculum focused on agroecology.

Thus, the incorporation of agroecology into the MST's agenda led to the creation of several Agroecology Schools in different states. There are currently about 30 Agroecology Schools in Brazil, mostly concentrated in the southern region of Brazil, mainly in the State of Paraná. In Paraná, there are five MST training centres and/or schools that present the debate and training for agroecology: Iraci Salete Strozak School, in the municipality of Laranjeiras do Sul and Ireno Alves dos Santos School, in Rio Bonito do Iguaçu, both connected to the Centre for Sustainable Development and Training in Agroecology (CEAGRO); José Gomes da Silva School, located in São Miguel do Iguaçu; Milton Santos School in Maringá; and the Latin American Agroecology School, in the municipality of Lapa. In these spaces, Technical Courses in Agroecology are offered in the following modalities: Integrated High School Technician in Agroecology Course; Technician in Agroecology – Youth and Adult Education; Technician in Agriculture with an emphasis on Agroecology; Technologist in Agroecology; Technician in Agroecology with an emphasis on Agroforestry Systems; and Technician in Agroecology with Qualification for Milk Production. The courses are held in partnership with public educational institutions, with resources from the National Education Program in Agrarian Reform (PRONERA), certified by the Federal Institute of Paraná (IFPR) and the National Institute of Colonization and Agrarian Reform (INCRA) (Novaes, SANTOS, 2014).

A document crafted by the Paraná chapter of the MST presents the main objectives of the Movement's schools in the State of Paraná.

- To be a training space for working-class organizations;
- To be a space where meetings of the MST and other organizations which seek the same goals of social transformation can take place;

---

the processing and commercialization of ecological grains with the trademark of Alimento Sagrado (Luzzi, 2007).

- To be a reference in the development of experiments in the field of agroecological production, presenting concrete results for farmers;
- To be a space for the development of socialist humanist values, matured through collective life;
- To improve methods of technical and political training and schooling since elementary school, as well as in high school and higher education;
- To enable the development of scientific and technological experiments, focused on the peasants' reality;
- To be a space where popular culture is encouraged and experienced, especially peasant culture;
- To be a space where people can live, educate, work, have fun, and build perspectives of the future

MST-PR, 2004; LIMA, 2011, p. 87

According to Lima (2012), the theoretical and methodological foundations of the MST Agroecology School in Paraná are linked to the Movement's philosophical and pedagogical principles of education and pedagogy, the systematization of which resulted from reflections on its educational-political praxis, based on three fundamental sources: socialist pedagogy, popular education, and dialectical historical materialism (SANTOS, 2015). Lima et al. (2012, p. 4) also present the Pedagogical Proposal of these schools, including the Milton Santos School, highlighting that it is

> based on the accumulation of work and experience in the training of militants and cadres in the MST, which is based on elements of the Political Pedagogical Project of the Florestan Fernandes National School (ENFF) and the educational experiences carried out at the Josué de Castro Institute of Education (IEJC). For the training of the Militant-Technical-Educator in Agroecology, it is necessary to appropriate the MST's political lines and organizational principles alongside technical-scientific, political, and organizational knowledge.

We observed that, in the pedagogy of the MST, education has a political commitment to social transformation, in line with the organization's educational practices. Thus, the training of the landless militants and educational practices in agroecology are not limited to activities developed in its schools, but are also forged in the matrices of human formation, among them, "the educational principle of work, social praxis, and history" (Caldart, 2004, p.42). Lima (2011, p.76) adds that

Educational practices in Agroecology have the political and pedagogical intentionality of an emancipatory educational project that is associated with the right to school and technical education in the rural reality. The understanding is that, in the settlements conquered by Landless families, collective alternatives can be built within the MST's organizational political praxis that inhibit the reproduction of capitalist social relations.

From the principles and actions of the Movement's agroecological education, *seeds* of non-capitalist relations can be born. Lima et al. (2012, p.9) state that "the organic structure and the management process, through the self-organization of the subjects, is the basis that enables the planning, organization and execution of the work, which is developed through cooperative relationships". Thus, agroecology is understood in the MST as an educational principle that guides the subjects to the construction of a societal project beyond capital. Roseli Caldart (2013) points out that today, agroecology is the movement's productive matrix. Along with cooperation, it serves as a guiding principle for all MST schools.

5 Agroecology in the Curriculum of the MST Vocational Schools

In the same way as in other types of MST-run or -influenced schools, in agroecology schools both the school organization and the curricula are implemented in a heterogeneous way. The MST Agroecology Schools tend to have an organic curriculum and in the state schools in which the Movement has influence, this agenda may appear in a less substantive way.

Agroecology courses, as well as those of a similar nature, are not only technical, typical of the hard sciences, with a large workload in the exact and biological disciplines. Especially because no course is just *technical*, not even the lauded state-run technical courses are neutral and free of ideology. Each and every course is political, and every technique carries with it politics, ideology, and class interests (Novaes, 2012; Dagnino, 2008).

The Agroecology School courses are not different in this way, as they disseminate a certain ideology in the technique they disseminate. They intend to carry out a critique of the Green Revolution and, more generally, the sociometabolism of capital. At the same time, they criticize the Green Revolution by disseminating research, documents critical of destructive production, and criticism of patents, pesticides, transgenics, etc. The MST courses contribute to the construction of theory and practice, and to the dissemination of the principles and foundations of agroecology.

While the courses of the State and Federal Technical Schools are conservative, with many *technical* disciplines that do not question the relationship between Science, Technology, and Social Classes (Dagnino, 2008), they offer little humanistic content and are based on a pedagogical project aimed at training the workforce within the paradigm of flexible skills/accumulation; in the MST Agroecology Schools, despite some contradictions and limits, it is possible to find a curriculum with more integration between social and hard sciences, an attempt to appropriate historical materialism and a fruitful relationship between theory, represented by the scientific foundations of agroecology and scientific criticism of the Green Revolution, and practice, through agroecological experimentation.

However, there are also difficulties in the MST Agroecology Schools. Guhur et al. (2012) observed that theoretical appropriation by the students is still low. In addition, Agroecology Schools face many contingencies: not all teachers are organic to the MST, there are problems in integrating disciplines, with relative fragmentation of knowledge, few resources for school maintenance, few researchers from Universities and Research Institutes developing agroecological research, and others.

6   Final Thoughts

Although agroecology for the MST is still an area under construction, we observed that, for them, agroecology is not reduced to preserving the environment. Quite the opposite: the Movement has tried to incorporate other elements and dimensions, such as: cooperation; gender issues; decommodification, and the rescue and systematization of knowledge accumulated by peasants, etc. At the same time, the MST Agroecology Schools challenge us to theorize radical criticism of destructive production and alienated labour. Furthermore, they challenge us to think about ways to bring agroecological theory and practice together, experimentations with the democratic management of schools and the educational systems of social movements. They also help us think about creating a curriculum that is both critical and more comprehensive, besides the classic divisions between social and natural sciences.

Agroecology, as a field of comprehensive knowledge which broadens the horizons beyond the epistemological homogeneity of uncritical positivism and postmodernist rejection of any epistemology, seeks to apprehend the historical character of social phenomena and the partial identity between the subject and the object through its practice. That is, theory and practice. However, agroecology, beyond a specific field of knowledge, is also a way of life that aims at an epistemological rupture with the fragmentation of reality and the destructive scientific utilitarianism of the so-called "Green Revolution". In addition to "sustainable development" or variations of "eco-capitalism", the agroecology of social movements helps us to think about agrarian reform, alternative forms of production, commercialization, and education beyond capital in the 21st century.

CHAPTER 8

Cooperation and Workers' Cooperatives of MST

1 Introduction

Since the 1970s we have been witnessing an offensive by capital. In Novaes et al. (2015), we outlined the main dimensions of this offensive: a) pressure for the free movement of financial capital, resulting in a restructuring of production in the countryside and cities; b) technological innovations that intensified the production and diversification of goods; c) expansion of capital towards sectors and fields not yet subject to full commodification, such as health and education, with a wave of privatization, which took the cycle of commodification to a new level; d) implementation of processes of corporate relocation and outsourcing.

The State of São Paulo is a privileged vantage points to observe this offensive in the Brazilian federation, especially the restructuring of the countryside and the (im)possibility of an agrarian reform that is based on cooperation, work in cooperatives and associations and agroecology in the settlements.

This chapter is divided into two parts. The first part makes a brief historical overview of the political economy of the countryside in the State of São Paulo. The second part shows the limits and contradictions of workers' cooperation and cooperatives in the MST settlements in the face of the new agribusiness offensive in the State. We close the chapter with some final considerations. It should be noted that the chapter was written for young people who intend to become Agroecology technicians. We seek to simplify some concepts and analyses without, however, falling into oversimplifications. We would also like to highlight that this is a part of ongoing research that aims to analyse the characteristics, nature, and contradictions of the workers' cooperatives of MST-São Paulo. The chapter we now present is the partial result of research on the actions of the capitalist state along with cooperatives and associations.

2 From the Coffee Complex to the Expansion of Agribusiness in the State of São Paulo

The State of São Paulo became prominent in the national scenario in the nineteenth century, with the consolidation of the coffee complex and a nascent industry (Cano, 2005). In this process, the importation of Italian and Spanish labour became fundamental (Martins, 2013). Not coincidentally, 15 days after the abolition of the transatlantic slave trade, the Brazilian oligarchies proclaimed the land law in 1850.[12]

The consolidation of the coffee complex led to the expansion of railway infrastructure in the State of São Paulo, with three main axes of expansion. To illustrate, we could mention the region around the cities of Bauru, Marília, and Presidente Prudente. In the twentieth century, we witnessed the wholesale slaughter of indigenous populations, the clearing of lands, the expulsion of squatters, land grabbing and the installation of large properties for the production of coffee and cotton (Monbeig, 1984, Pereira, 2005). José Teodoro, a miner from Pouso Alegre, bought the lands in the region, leading to urban agglomerations that grew as railroads arrived in the region.

With the stock market crisis of 1929 and the Revolution of 1930, we entered another historical moment. Wilson Cano observed the "roots of industrial concentration" in the State of São Paulo. This is where the country's main banks, commerce, and industries emerged. The emergence of the industry– late as it was–led to the creation of the general conditions for its production and reproduction. It led to the creation of a timid educational system, the creation of Senai/SESI and the Federal Technical Schools, the creation of USP and the Isolated Colleges in the countryside (Saviani, 2008). Along with these, a housing policy focused on workers' villages and on controlling rental prices was created (Bonduki, 1999). Francisco de Oliveira (2004) also highlights the exchange rate, subsidies, and numerous state policies favourable to industrialization. With that, the resources coming from coffee production were "drained" for the consolidation of the industrial park. Carone (1981), in turn, emphasizes the attachment of trade unions to the State as a way of avoiding and framing the demands of the nascent proletariat within the framework of capital.

From a political point of view, Ianni (2009) notes that the São Paulo elite reacted quickly to the loss of state control to Getúlio Vargas. The Revolution of 1932 was an attempt of certain classes from São Paulo to regain control of the State. Also according to Ianni (2009), from 1930 to 1955 we had a brief moment

---

1 Clovis Moura (2014) also observes that in the second half of the 19th century struggles against slavery were already at a new level, and this changed the history of black people in the State of São Paulo. In São Paulo, it is worth highlighting the history of the abolitionist Luiz Gama. Born in Bahia, he had been enslaved, and was a poet and lawyer. He lived for 42 years in São Paulo and fought for the end of the monarchy and the abolition of slave labour in the State of São Paulo. He died in 1882, 6 years before the abolition of slave labour. See Moura (2014).
2 T.N.: The Land Law of 1850 highly favored the concentration of lands in the hands of a few wealthy owners, consolidating Brazil's still prevailing inequality in land distribution.

when autonomous capitalism flourished, but quickly became an associated capitalism, with decision-making centres outside the country. For him, foreign capital began to determine the direction of the nation, something that became clearer from the 1964 coup.

Let us remember that in the period between 1950 and 1964, class struggle intensified throughout the country. Struggles of the nascent proletariat for better wages and living conditions in a country with uncountable scars of slavery, struggles for land in a country dominated by large landowners, urban reform in a country that has replaced slave quarters with favelas, and the creation and reform of universities linked to national development.

Struggles to expand public and secular education, such as those led by Florestan Fernandes and his friends and literacy campaigns, especially those influenced by Paulo Freire. Campaigns such as "The Oil is Ours", struggles for the creation of the Eletrobrás system, among many others. Glauber Rocha became the icon of the "Cinema Novo" (New Film) movement, Sérgio Ferro of the New Architecture, the students of Nelson Werneck Sodré of New History. These struggles were part of the most beautiful and delicate moment in our history.

An alliance between the USA, the industrial bourgeoisie, the Catholic Church, large landowners, and businessmen, with prominent participation by the São Paulo elite, led to the overthrow of João Goulart on March 31 to April 1, 1964. We also had numerous street demonstrations including the march for "Tradition, Family and Property" and campaigns inciting hatred for "communists", among others (Dreifuss, 1981).

Marxist historians are unanimous in observing that the civil-military dictatorship represented a major historical break. Leftist professors and students in USP, UnB and many other federal universities were decimated, struggles for land were put to a stop, left-wing parties and unions were made illegal, many influential figures were exiled, critical thinkers was tortured and killed, the leftist part of the Catholic Church was strangled, and the peasant leagues disintegrated. (Fernandes, 2006; Novaes, 2012).

The restructuring of São Paulo's countryside, that is, what Florestan Fernandes called its "consented modernization", led to the installation or expansion of large corporations producing pesticides, synthetic fertilizers, and tractors. For the "modernization" of the countryside to consolidate, it took a heavy hand from the State to create the general conditions for its expansion: very cheap credit for the purchase of machinery and equipment, technical assistance to implement the "Green Revolution" package, financing exports, repression of the countryside unions, extermination of guerrillas, reform or creation of public universities to adapt the curriculum to the dictates of the "Green Revolution", etc.[3]

Within this history, we could highlight the march of the gauchos[4] towards the west of Santa Catarina, Paraná, Mato Grosso do Sul, Mato Grosso, and Pará. They've even settled in Bolivia. Murders of indigenous people and squatters, land grabbing, and control of regional politics by large-scale farmers all reared their ugly head in our history yet again. Among the debate on the restructuring of the countryside, Nelson Werneck Sodré (1995) highlights the contradictions of Proálcool. For him, if on the one hand Proálcool meant at that time the production of national energy, on the other it led to the concentration of rural property and income. At the end of the day, this policy led to the formation of groups of large landowners, with an enormous power to decide the course of Brazilian capitalism.

The explosion of numerous demonstrations and struggles starting in the second half of the 1970s led to the strategy by part of our elites of a gradual, slow, and secure transition out of the dictatorship. This strategy was successful, leading to a transition from dictatorship to "democracy" without major ruptures, a democracy increasingly dominated by financial capital. The control of the strategic positions of the State by capital limited the power of the so-called "new unionism", of the PT and PMDB, Central Worker's Union (CUT), the MST, ANDES, and other institutions that tried to recompose the classic social struggles of a country torn by numerous social contradictions. When we thought we would win this safe transition, with the 1988 Constitution as our high point, the media conglomerate Globo elected Collor who, together with the increasingly transnational and restructured corporations, knocked the people out once again. Most of the achievements formally inserted in the Constitution were denied to Brazilian workers in practice. To give you an idea, in the 1980s the Pontal do Paranapanema became one of the largest stages of struggle for land in Brazil. On one side we had lands grabbed by large landowners and corporations with the connivance of the State and on the other side the landless movement.

In the 2000s, sugarcane crops, which already occupied 50% of the territory of the State of São Paulo, grew to occupy 80%. The State of São Paulo went from being a large coffee plantation to becoming a large sugarcane plantation. Sugarcane has become a kind of "transparent" gold. Let us remember Lula said in 2005 that sugarcane would be the salvation of the Brazilian countryside. To

---

3 For this debate, see Delgado (1985), Tolentino (2011), Sampaio Jr. (2013) and Novaes (2012).
4 T.N.: Natives of the southern state of Rio Grande do Sul.

give you an idea, the profitability of the "transparent gold" led to the installation of over 80 sugar processing plants in São Paulo, Goiás, the Triângulo Mineiro, and Mato Grosso do Sul.

It should also be noted that if until the 1970s the struggle was against "top hat farmers" (as a member of the MST once said), from that point on on the struggle is against farmers and transnational corporations that invest mainly in the sugarcane fuel and eucalyptus market, buying or controlling São Paulo's lands and placing local class struggle on another, much more complex level.

The productive-economic control of agribusiness will manifest itself in the reproduction of the political-ideological apparatus. Regarding the Judiciary Power, there are many studies demonstrating the variety of maneuvers to delay the creation of landless settlements, question existing settlements and using every legal means to hinder and block the creation of settlements.

In education, Lamosa (2013) draws attention to Agribusiness Education in the schools of Ribeirão Preto. We believe that currently an education focused on "sustainable development" has gained strength, placing itself next to pedagogies based on "affection", "tolerance" and "entrepreneurship", widely disseminated by the State Department of Education São Paulo, as part of the new dictionary of capital.

In a preliminary analysis of the Brazilian High School Curriculum, conceived by capital's organic intellectuals, one can perceive the absence of a radical criticism of destructive production, that is, an analysis that is not adequate for the perception and fight against transgenics, pesticides, synthetic fertilizers, and their fundamental pillars – the control of land by capital. Quite the opposite. The student and teacher textbooks talk about knowledge societies and the evolution of science, propagating an idea of a simple "choice" between organic and transgenic products hovering to young workers in public schools. "You decide", as long as you don't fight the landed corporations and the landowners, let alone the "Green Revolution" package.

One particularity of the State of São Paulo is that the same party – PSDB – has been running the state in coalitions since 1994. The so-called São Paulo Social-Democrats promoted a broad counter-reform of the State, based on privatizations, commodification of education and readjustment of the higher education and research complex, reform of technical schools and an expansion of mass incarceration, public-private partnerships and social organizations (Sanfelice, 2010; Novaes, 2014).

We have teachers asking for "less bullets and more chalk", homeless people fighting for "more rights and less right-wing", and people affected by dams saying "water for life and not for death". As the fractions of capital that command the State of São Paulo no longer have republican solutions on their horizon, the recurring solution is police agression (Novaes, 2014).

In our view, it is within this context that the scarcity of policies for land redistribution and settlements, in contrast to plentiful policies for financialized corporations, must be analysed – that is, as a project to create *obstacles* for agrarian reform and the annihilate anti-capital social movements. For hard-won agrarian reform settlements, we have sparse, slow, inadequate, disconnected state policies, diffusion of a utilitarian vision of cooperatives and associations by the Federal and State Government, and an exotic view of agroecology, in which it isn't the fundamental pillar of agrarian policy.

For the corporations, we have "innovation" policies, policies for the expansion and installation of sugar and alcohol plants, industrial parks and hubs, real estate speculation in strategic regions for financial capital, such as Campinas, Greater São Paulo, Ribeirão Preto, Araçatuba, and the Paraiba Valley. Given this scenario of aggressive capital, it is unlikely that workers' cooperatives and agroecology can flourish, as we will see in the section below.

3   Conceiving Cooperation in the MST

The struggles and resistance to slave labour and against the ills of land monopoly are part of our history (Moura, 2010). Several experiences of resistance such as Quilombo dos Palmares (destroyed in 1695), Canudos (1887), the Cabanagem (1840), Contestado (1916) and the Peasant Leagues (1947 to 1964) were harshly repressed by the Brazilian State representing a pact of elites and landowners with foreign capital.

In the region of Ribeirão Preto, for instance, the Peasant League of Dumont is created with the help of the Brazilian Communist Party (PCB). It was led by João Guerreiro Filho, Pedro Salla, Miguel Bernard and Vitório Negre, militants defending the interests of the peasants. In 1947, after the PCB became illegal, the home of the Guerreiro Family was invaded and searched by state forces (Welch, 2010, p. 131–32). Welch (2010) notes that the PCB's rise in popularity between 1945 and 1947 and its electoral success led the government to make the party clandestine in May 1947; the peasant leagues were gradually repressed. Welch continues:

> On May 9, the police raided the party at the UGT (General Workers' Union) headquarters and confiscated a long list of political apparatuses and organizational material, including six red invitations for the "Rural Workers of Barrinha", a statute for "Agricultural Salaried Workers" and

two monthly receipts for members of the Peasant League of Dumont. Within a day, the police raided Guerreiro's parents' home and took away the league materials he had left there. Six police officers ransacked the house of peasant Pedro Salla, but no documents were found.

WELCH, 2010, p. 145[5]

The emergence of the MST and other social movements that contributed to the process of "redemocratization" of the country in the 1980s, through the direct action of workers (with the tactic of occupying unproductive land in large estates) shows the struggle for land was still latent in the scenario of the Brazilian State, which despite harshly repressing these struggles, failed to contain the reorganization of the working class in the countryside.[6]

This strong stance of struggle and land occupation by organized workers, particularly in the MST, forged the creation of several agrarian reform settlements. The territories, acquired under the organizational influence of the MST, fought to maintain and progress in a more or less independent way, not subordinated to agribusiness and big capital.

Considering that the settlements are immersed within a broader space of capitalist nature, the settled territories were not and are not immune to capital's offensive and a possible reconcentration of the settlements if no alternative socio-productive organization action was taken (Christoffoli, 2012).

After the first settlements were established, it became increasingly clear to the militancy of the MST that the struggle for land could not be limited only to establishing settlements, as capital would not give up its lands without resistance, without trying to convert them to its operating logic, when not directly reclaiming that lost territory. We highlight the process of commodiffication of the countryside, encouraged via the Green Revolution and all its stages during the military dictatorship, which had a great impact on the productive organization of Brazil. Landless workers were not immune to this process of dependent industrialization and commodification of the Brazilian countryside.

Within this context, the Movement saw that in addition to the occupation and conquest of territories, there was a need to organize and enhance production in established settlement areas. The stance taken was to "develop

---

5 "In September 1949, in the municipality of Tupã, a meeting of militants from the Brazilian Communist Party (PCB) that was holding a Congress of rural workers in Alta Paulista was violently repressed by the police. Three communists and a policeman died in the clash. Despite the great national repercussion at the time, the episode was relegated to oblivion in the history of the city" (Lima, 2009).
6 See Martins (2004), Stedile (1999) and Rodrigues (2013).

cooperation as a type of strategic action in view of the advance of capital over redistributed areas, but also as a trial for the future organization of agriculture in a socialist society" (Christoffoli, 2012, p. 171).

It is noteworthy that at first – between 1979 and 1983 – the organization of the struggle for land received great influence from the Catholic Church. In this dynamic a logic that we can call "Land of God, land of brothers" was disseminated. Productive organization was thought of less in the perspective of economic results, but more in a perspective of divinity linked to the logic of the church (Mateus, 2015).[7]

Organized in family units with small productive scale, low intensity of capital use, labour-intensive technologies, "low" development of the productive forces and almost no insertion in the markets, it was a subsistence economy complemented by the insertion in the market of scarce lines of credit (Christoffoli, 2012, p. 172).[8]

The concept was that everyone should work together, sharing what they have, and to have subsistence as a goal. In this dynamic, small associations were created, not with an economic objective, but with the aim of building a Catholic peasant community. However, the 1980s saw a growing mass movement in the countryside happening alongside an opening process in the tail end of the Military Dictatorship, bringing the MST to political involvement, with struggles, occupations and achievements that transcended the concepts of the church, leading the MST to discuss its own concept of cooperation (Matheus, 2015).

Christoffoli (2012) points out that the first experiences of associated work, despite being vital for the survival of workers, failed to reach high levels of coverage. Many initiatives were made unfeasible by the difficulties imposed by the state bureaucracy and mainly by the logic of capitalist production (access to capital, technologies and management).

The difficulties faced in this first moment of the associative and/or cooperative workers' organizations, whether due to the limits of the workers' own concepts or due to the land regularization process carried out by the State without considering the economic viability and support of the families, led to the constitution of impoverished settlements, causing many workers to seek external employment to supplement their income.

Considering the experiences of this first moment, which was based on collective and semi-collective groups, and later on studies of the experiences

---

7 See also Stedile and Fernandes (2005).
8 See also Cerioli and Martins (1999) and Martins (2004).

of socialist countries (Cuba, China, East Germany, Bulgaria), a reference of cooperative organization with a "socialist" nature begins to be developed in the MST. Based on the collective use of the means of production and labour, it begins through small collective groups that later become large fully collective cooperatives. A higher level of organization of cooperatives encompassing entire settlements are called Agricultural Production Cooperatives (CPA).

João Bernardo (2012) describes that, faced with a scenario where rural policies penalized small farmers, the MST focused its efforts on building cooperatives. It was necessary to find means and conditions that would allow workers to have access to financial and technical resources and favourable conditions for production and distribution.

The leap in quality in the Movement's concept would be to socialize everything, all the "factors of production": land, capital and work. In this scenario, the MST enters a new phase, based on the Cuban experience: the Movement evolves in debate and action, thinking about the concept of a cooperative system for Brazil, a national system with the function of meeting the demands of settlers in different realities throughout the country.

The organization of cooperatives associated with the consolidation of agro-industries had the objective of inserting agrarian reform products in the market. It was believed that through these actions there would be a qualification of production and consequently social and economic development for settled families.

It is noteworthy that at that time the MST was fighting for a classical type of agrarian reform. For Toná (2011), the Movement believed that there was an interest of the ruling class in playing along, and consequently the possibility of insertion of peasants in capitalist production, with complementary interests between these and the industry.

In the early 1980s the number of agricultural cooperatives in the Movement's settlements had increased considerably. In the late 1980s, the MST created the Cooperative System for Settlers (SCA). This process led to the formation of the National Confederation of Agrarian Reform Cooperatives in Brazil (Concrab) in 1992, from approximately 55 production and marketing cooperatives and 7 state cooperative centres. Along with this, over 40 CPAs were organized, "many entirely collectivist, true socialist islands not only in terms of the organization of work, but also in terms of certain aspects of domestic life, such as the use of cafeterias and day-care centres" (Bernardo, 2012).

The production cooperatives founded by the MST had legal registrations so that they could enter the market. Their organization seeks the autonomy of the settlements in addition to production planning and the creation of direct commercialization routes, eliminating traditional middlemen. In other words, the idea was to have control over the entire production chain (Bernardo, 2012).

Despite the euphoria, the challenges and contradictions imposed by the capitalist logic were huge:

> the lack of knowledge and peasant distrust of these collective forms resulted in a partial reversal of experiences, initially breaking up into semi-collective groups and finally in the complete dismantling of several complex cooperation initiatives. The gap between the proposal conceived by the movement, of total self-managed collectives, with the lack of state support, insufficient technical preparation and the contradictions derived from the artisanal organizational consciousness of the peasants were fatal for many of these experiences and forced a tactical retreat from the movement.
> CHRISTOFFOLI, 2012, p. 175

It cannot be ignored that, at this juncture, state action itself posed several obstacles and difficulties to materialize the Movement's proposals. The lack of understanding and use of institutional legal apparatus to enable self-management and the collectivization of the means of production and work acted as strong inhibitory agents and using them ran against the MST proposal.[9]

In view of this, mainly through Concrab, which started to focus on actions related to cooperation from 1994 onwards, efforts began to be directed towards the constitution of regional cooperatives for the provision of services and no longer on self-managed collectives. This model allowed greater flexibility for the organization and agglutination of settled families who produce individually on their lots.

In the midst of this context, the concept of cooperation in the MST seeks to transcend the simple issue of production or even bureaucratic legal organization. It must also explore the potential of an ideological political formation

---

9 The Collor government, for example, restricted credit and technical assistance for small-scale agriculture, extinguished the Ministry of Agrarian Reform and Development, emptied the National Institute of Colonization and Agrarian Reform, Incra, and appealed to the Federal Police to repress the MST, ordering state secretariats to be invaded, seizing documents and arresting and instituting legal proceedings against its leaders. This period of repression caused a drop of almost 50% in the total number of land occupations, which dropped from 80 in 1989 to 49 in 1990, and practically half the number of mobilized families, which dropped from 16,030 to 8,234 in the same period. In this difficult context, having to survive defensively and relying mainly on its own resources, the MST focused on the development of production cooperatives (Bernardo, 2012).

necessary for the settlers to participate in struggles and solidarity claims to other categories beyond field workers.

For Christoffoli (2012, p. 55), currently, the MST's concept of cooperation has a perspective that joins economic and social development by developing humanist and socialist values. He believes that it should not be restricted only to organizational, political and economic objectives: it should also be understood as a tool of struggle built collectively so that it should contribute to the organization of workers settled in grassroots nuclei, as well as the practical training of militants (workers) for the political, economic and cultural struggle.

4   Cooperation of the São Paulo MST in Face of the Capitalist State

As highlighted in the previous topic, during the 1980s land occupations grew as a tactic to force the agrarian reform process. In the State of São Paulo – during the Montoro government (1983–87) – there were 3 occupations, in the regions of Itapeva and Campinas/Sumaré and in Pontal do Paranapanema. At the time, the Institute of Land Actions (IAF) was created, a state agency that established dialogue with the MST and other social movements in the countryside. Today the MST has around 150 settlements in the State of São Paulo.

In parallel with the transformations of the MST as a whole, the internal discussion of the Movement was already seeking to overcome those first associative and/or cooperative concepts and organization models founded on the perspective of the church, with an eye on creating cooperatives or associations to improve production and market access with the goal of improving the income of settled families.

Besides creating the IAF, other instruments that no longer exist were also created during the Montoro government, such as credit lines and technical assistance teams for the settlements. The perspective was to move forward with the agrarian reform in the state of São Paulo and, to do so, machines were financed through Caixa Econômica Estadual[10] to enable the creation of associations of machines and services.

The experience of a settlement in the city of Pirituba can be used as a reference. Unlike the current process, the settlement was not divided into individual lots, but organized into collective plots made up of between 10 and 15 families that received support and assistance from the machines and services

---

10   The Caixa Econômica do Estado de São Paulo was a bank run by the State of São Paulo from 1916 to 2009, when it was incorporated to the federally managed Bank of Brasil (BB).

association. In this sense, the Movement's own concept of cooperation was already under development, aiming to create a cooperative system in Brazil, a national system that sought a certain autonomy of the Movement alongside control and qualification of the production chain.

With the change from the Montoro government to the Quércia government, problems arose for the advancement of agrarian reform in the state, directly affecting the discontinuity of the new cooperative system. It's not that problems were exclusively due to the change in the state government, but rather we're pointing out that it had an important influence on the process. With the new administration, a movement against the dialogue of the government with working class social organizations took hold, causing the replacement of the entire staff of the agency in charge with a new team that was opposed to the MST in the State of São Paulo, indirectly fighting the Movement's experience which excelled in self-management and collective organization of the settlements.

Both in the state and in the federal scenario, with the start of the Collor de Melo government and the adoption of neoliberal policies, alongside the Movement's limits in internal capacity building, the CPAs fell into disarray, many of them succumbing definitively. Of the few remaining in the state of São Paulo, Coopava in Itapeva stands out (Santos, 2015). In this context, at the national level the MST resorted to a posture aiming to avoid being completely overwhelmed by the crisis. To this end, regional cooperatives were created to provide services and commercialization, in order to meet the demands and strengthen the organization of the settlers regionally.

If we look at the advances and setbacks of the MST, it is possible to see that there is an oscillation directly linked to the way governments repress, react and anticipate the actions of anti-capital social movements. At times when the agrarian reform debate advanced and the government, to some extent, gave in, that is, when they created some mechanisms to actually advance agrarian reform, the experiences advanced. In periods when the government reversed course, however, the experiments went into crisis.

As an example, we highlight the federal government of Fernando Henrique, which sought to conduct the economic aspects of agrarian reform, in order to isolate the MST politically and dismantle its social base. To this end, in 1998 it created the Banco da Terra, with the aim of replacing occupations with access to land through market mechanisms.

João Bernardo (2012) points out that the most striking strategy taken by Fernando Henrique Cardoso (1995–2002) in his confrontation with the MST consisted of supporting family farming and promoting the direct relationship of peasant families with the market, to the detriment of the collective relationship carried out through settler cooperatives.

This action resulted in the blocking of the Special Credit Program for Agrarian Reform, Procera, which was extinguished in 1999, but which was replaced in 1995 by the National Program for the Strengthening of Family Agriculture, Pronaf, whose name is illustrative. "The following year, Pronaf ceased to be just a credit line and became a government program. It was about dismantling production cooperatives, diverting credit to family farming" (Bernardo, 2012).

The replacement of Procera by Pronaf put the CPAs in huge financial trouble and in a way forced the MST to stop privileging the formation of cooperatives with a more complex scope, instead presenting cooperation proposals linked to conventional forms of commercialization that do not guide work processes. Since then, the MST management has given priority to service provision cooperatives.

By initiating this new line of promotion of family farming through Pronaf credits, Fernando Henrique Cardoso (1995–2002) achieved a notable strategic triumph, in a presidency that was otherwise not marked by great successes. As in so many other aspects, the Lula government extended the guidelines of the previous government. Pronaf funds quadrupled between the 2002–2003 and 2006–2007 harvests (Bernardo, 2012).

5   Cooperation and Cooperativism in the MST of São Paulo

We could illustrate our arguments with the experience of the Reunidas Settlement (Promissão), of COCAMP in Pontal do Paranapanema and of the Sepé Tiarajú Settlement.

According to Santos (2007), Fazenda Reunidas was the centre of the social dispute over land in the region. In 1983, the possibility of expropriating the farm for Agrarian Reform began to be considered, which eventually materialized with the First National Plan for Agrarian Reform. The Reunidas Settlement is located in the municipality of Promissão, in the countryside of the state of São Paulo, 450 km from the capital. It appeared in the mid-1980s, contemporary with the reformulations of the National Plan for Agrarian Reform (PNRA). Local discussions on agrarian reform and democracy resulted in the formulation of a concrete demand: the expropriation of Fazenda Reunidas. In June 1985, Fazenda Reunidas appears in the list of properties that could be considered a priority for Agrarian Reform in Brazil.

Also according to Santos (2007), on June 30, 1986, President Sarney signed Decree-Law No. 92,876 expropriating Fazenda Reunidas. In 1987, a Declaration of Possession was made by the Union and the registration and accommodation of families began. At first, 800 were selected, alongside a group from Campinas, which had camped with 350 families, and a Group of 44 families who had been camping there for over a year. Soon after, the Group of 44 received authorization to use the farm's land for their specific purposes.

The workers who occupied Fazenda Reunidas came from several cities, such as: Lins, Getulina, Promissão, Ubarana, José Bonifácio, Sabino, Birigui, Penápolis, and Campinas and its surrounding regions. These workers lived different histories, marked by the experience of salaried work, sometimes temporary and sometimes permanent, or even by the experience of being sharecroppers, tenants and partners (Santos, 2007).

In June 1988, some selected families began to be settled, grouped by their municipalities of origin. In this phase, most of the newly settled families built their canvas or mud shacks divided into agro-villages (agrovila in Portuguese), a way found to organize the settlement socially. These agro-villages were organized by the workers' region of origin, thus, the distribution of families was as follows: 101 families belonging to the Agrovila dos 44; 78 families to Agrovila Birigui; 98 to Agrovila Lins or Central; 12 families belonging to the Grupos dos Doze; 80 families to Agrovila José Bonifácio; 74 to Agrovila de Campinas; 83 to Agrovila Penápolis; 31 families located in the Agrovila do Cintra; 30 to Agrovila São João; and finally 42 families belonging to Agrovila São Pedro. A total of 629 families distributed in 8 agrovillages were settled in 19.36 hectares each, except for Agrovila de Campinas, where, due specificities in the land occupation process, each family was entitled to 17 hectares of land on average.

Santos (2007) divides the constitution of the Reunidas settlement into three moments. The first results from the organized action of workers near Promissão in 1986 with guidance from the Pastoral Land Commission – CPT, when 44 families camped on the edge of Fazenda Reunidas, aiming to pressure the government to expropriate the land. In the following moment, in 1987, after the farm was taken over by the Union (October 29, 1987), three hundred and fifty families from the region of Campinas/SP arrived at Fazenda Reunidas, on the edge of BR 153, and formed another camp, called Padre Josimo Moraes de Tavares. These families were organized by the Landless Rural Workers Movement – MST and by a group of the Ecclesiastical Base Communities (CEBs) of the region since 1985. And the last moment, in July 1988, in which the families selected by the selection commission are settled by the Federal Government, organized according to their cities of origin.

The foundation of the settlement's social organization were the agrovillages, which can be compared to neighbourhoods in urban areas, and are divided according to each family's city of origin. The agrovillages have an area set aside for houses, schools, a health centre, a community centre and leisure areas (Santos, 2007).

As for the process of the Reunidas Settlement's formation, there was no discussion or creation of a Program, project, or Settlement Development Plan with the settled families, nor was a diagnosis of the area of the Settlement Project prepared to come to proposals that contemplat ed a vision of the settlement's future with production programs, considering a production system to be implemented and economic feasibility analysis and observation of social aspects. Given the delay in the project's implementation, the lack of infrastructure and the lack of definition in the division of the lots, the first families called by the selection committee came to the site with the single aim of occupying the area. In the early 1990s, the vast majority of these settlers began to live in the settlement without their families. The peasants built their shacks and made small gardens that were tended only by them, with occasional help from family members.

Santos (2007) observes that there was still no financing for investment, nor were there resources to invest in any productive activity. The choice of productive activities was much more linked to the credit lines that were available than to the aptitude or economic viability of these activities.

Regarding the internal organization of settlers, the "tractor groups" stand out among the first organizing experience the settlers had. This initial organization resulted from the release of investment credit in the late 1980s, in which over 400 families had access to this financing for the purchase of tractors and implements that were distributed among groups of 10 families throughout the settlement. Hence the name: tractor groups. This experience of group work lasted for two harvests – after that, the groups were disarticulated, and since then, there has been a strong predominance of family work (Santos, 2007). For him, "what we see is that there is a cultural problem in relation to these organization practices. There is a predominance of individual property and a culture of the settler being his own boss, so settlers see the formation of an organization as a loss of freedom to produce and grow with their family" (Santos, 2007).

Also according to the author, the experience of the Padre Josimo Tavares Agricultural Production Cooperative – COPAJOTA was the one that most contributed to the development of workers' organizations in the Reunidas Settlement. Since then, four more organizations have emerged.[11]

---

11   According to Santos (2007), "COPAJOTA represented several advances for its period, 1992 to 1998, but it was also rife with serious internal organization, management and political problems. The internal organization model was based on the experiences from the South of the country, mainly Rio Grande do Sul. The distribution of surplus was through hours

In 1995, COPAJOTA had 15 greenhouses growing lettuce, tomatoes, bell peppers, cucumbers and green beans, with a weekly production of over 150 boxes; in dairy farming, there were 90 heads of cattle with a daily production of 400 litres of milk. The advance of surplus gains happened monthly and the value per family was of around two minimum wages (R$ 200.00, at the time), not counting production for self-consumption. The only source of funding for new investments was funding from PROCERA and FEAP. This was reflected in 1998, when a high degree of indebtedness could be observed. The cooperative could no longer continue its productive activities, and backlogged payments for financing, loans and debts started to accumulate. In 1999, the cooperative stopped all its activities. It was only in 2004 that the process of individualization and renegotiation of PROCERA, FEAP and Finsocial debts was resumed in order to regularize the financial situation of its members. This experience of organization has an extremely important contribution, both in discussions about the different forms of worker organization, production models, planning and management models, types of distribution of profits, as well as in relation to their role in the development of the settlement (Santos, 2007).[12]

Ribas (2004) analysed the case of COCAMP in Pontal do Paranapanema. It was founded on December 28, 1994, by 291 partners at the headquarters of Fazenda São Bento, in Mirante do Paranapanema. Between the end of 1994 and throughout 1995, the cooperative focused exclusively on activities related to the organization of its documentation. Starting in April 1996, the cooperative started to have physical structures based on its agricultural and agro-industrial projects.

Ribas (2004) states that the genesis of COCAMP is linked to the territorialization process of the MST in Pontal do Paranapanema, since the settlements' founding, the movement began to connect the struggle for land with the need for political and territorial management of the settlements (organization of production, distribution, construction of houses, basic sanitation, etc.).

Ribas (2004), as well as almost all the other researchers mentioned here, observes that the first experiences on cooperative organization developed by the MST consisted of collective groups, mutual aid groups, based on the practices of the Catholic Church. Subsequently, they encouraged other

---

worked and not through capital contributions, that is, work prevailed over capital. In 1994, this form was strongly questioned by the associates, as they wanted the distribution of leftovers to be based on ownership of the lot, that is, divided in equal parts by family and not by the hours worked by the partners, resulting in the departure of 17 families and over 30 members, the first 'split' in the cooperative's history".

12 To learn more about the Reunidas settlement, see also Leandro (2002).

experiments with machine groups and associations. In the late 1980s, the MST began its experiments with small cooperatives, the Agricultural Production Cooperatives (CPA's), characterized by an entirely collective management of lots and production.

In this author's point of view, the MST leadership learned from these attempts that there was a growing need to enhance mass cooperation. This redefinition resulted in the creation of the Service Provision Cooperatives (CPS's) and from these new experiences discussions about regional cooperatives began. In 1993, COANOL, in Laranjeiras, in the state of Paraná, and COAGRI, in Sarandi, Rio Grande do Sul, were born. In 1994, in Pontal do Paranapanema, COCAMP was created, resulting from a series of discussions to assist groups of families from existing settlements (Gleba XV de Novembro, Água Sumida, Santa Rita, Che Guevara, São Bento and Rosana). COCAMP currently has 2,220 members who are distributed across 12 municipalities in the Pontal do Paranapanema region (Ribas, 2004).

The Sepé Tiarajú Settlement is located in the State of São Paulo between the municipalities of Serra Azul and Serrana-, near Ribeirão Preto. In April 2000, 100 families occupied an area of approximately 790 hectares of the former Usina Nova União, on land taken by the São Paulo State government from the mill owners, as payment of debts and social taxes.[13] In August 2003, INCRA bought the area, and on September 20th, 2004, the process of settling 80 families distributed in four agrovillages began as a PDS – Sustainable Development Project (Scopinho, 2006).

The presence of the MST in the area put organizational models for rural settlements into question. For the movement, a settlement becomes a space for the construction of new social relations insofar as it is structured from an organizational process that considers both the economic dimension and the

---

13   Scopinho (2006) also observes that the changes resulting from the internationalization of the economy accelerated the reconfiguration of the productive base, causing a strong impact on the supply of jobs in the state of São Paulo. In regions with an agro-industrial economy such as Ribeirão Preto, the intensification of agricultural mechanization through the use of mechanical harvesters caused structural unemployment in agriculture, especially in sugarcane-producing regions. Unemployment, both urban and rural, combined with the absence of alternatives for income generation and the State's omission regarding the situation of the unemployed, forced this population to survive precariously from informal work on the outskirts of the so-called "dormitory cities" as freelance rural workers (known as boias-frias) or to migrate seasonally to monoculture regions. The desire for social and political participation revealed the most unequivocal evidence of the possibility of creating roots, because it concerned the desire to actively participate not only in that collective, but also to help in the construction of others (Scopinho, 2006).

survival strategies of families and dimensions related to social interaction and political and community participation. This is important in order not to reproduce the social relations prevailing in the Brazilian rural world which, according to movement leaders, occur because the worker, by force of circumstances, organizes his time and his life around the family's immediate survival strategies, limiting their political participation in demanding wages and/or better prices for their products. From the MST's point of view, organized settlements can become spaces of economic and political resistance for rural workers (Scopinho, 2006).

At Sepé Tiarajú, the main challenge of the organizational process was how to converge sociocultural diversity in the sense of carrying out a settlement project, collectively built on the basis of cooperation and agroecology, with socioeconomic and environmental viability, without denying the traditions, customs, needs and interests of the settlers. For Scopinho (2006), coexistence and dialogue were important strategies to understand the characteristics, thinking and daily work and life of families.

In the settlement, since the camp days, production has always been "agroecological". Due to a lack of resources to buy inputs, seeds and large machines and under the guidance of the MST, the settlers had to rescue old techniques or invent ways of producing that did not require "advanced" technology. The biodiversity of plants and animals in the orchards and improvised gardens around the shacks has always attracted the attention of visitors, which gradually brought back some species of birds and small animals to Fazenda Santa Clara (Scopinho, 2006).

Among the main advantages pointed out about cooperation, Scopinho (2006) highlighted those related to improvements in the organization, in the planning system and in production management. The settlers saw the social advantages of cooperation, in the sense of facilitating social, political and cultural relations and improving living conditions. They also understood that the rational use of scarce resources reduces production costs, because together they can buy inputs and provide the necessary logistics to produce. Both idealized and abstract meanings were revealed, ranging from increasing the settlers' self-esteem to minimizing isolation (Scopinho, 2008).

Scopinho (2006) points out that for those who had no experience with rural work, cooperation also meant a space for coexistence that favoured training for work through learning cooperative and agroecological ways of working the land, creating the possibility of getting to know the people who live in the community, of discussing common problems and planning production and life with a smaller margin of error, for improvement of living conditions, for more dignity and social protection from the insertion and coexistence in an organized group, enabling discussions on political and social issues of interest to the settlers, such as the national political situation and the directions of agrarian reform, the causes of rural exodus, and the absence or inadequacy of public policies for rural populations.

Scopinho (2006) believes this learning would increase the chances of participation and interference in political decisions at the local, regional and national levels and would help in the creation and maintenance of organizational structures to enable cooperation and training. In short, for the settlers, cooperation could be a new path to new roots, contributing to create conditions for the economic, social and political survival of their communities.

Also according to this author, the contradictory meanings of cooperation were present in individual values and in the way the settlers saw their own condition in the relationships they established with each other in daily life at the settlement. In Sepé Tiarajú, many settlers saw cooperation as an imposition of the MST and INCRA and their responses were controlled by the fear of losing the opportunity to be settled. For this reason, we sought to learn what meanings were attributed by settlers to cooperation in terms of perceived advantages and disadvantages when experiencing it in the daily life of the settlement, mainly from the experience of working collectively.

In the settlement, some challenges arising from individual differences were mentioned, including the belief that cooperative work could be similar, in a hierarchical character, to the wage labour. The perception of these individual differences leads to a crucial issue in the cooperation ideology, which is the notion of justice in the distribution of work and its results. Two important issues are involved. The first concerns the idea of a lack of standardization tasks and worker behaviour, which is a characteristic requirement of heteromanagement to guarantee the reduction of production costs, the increase of work productivity and quality assurance of the product.

For this researcher, the principle of equality inherent to the cooperative ideology would have the disadvantage of not standardizing and not homogenizing. Consequently, the second issue refers to a perception that there was a lack of mechanisms for equitable distribution of the results of the work that, according to the settlers, would be related to a lack of rules non-compliance to them. At the same time that they felt autonomous and felt like the owners of their own mechanisms to eliminate social inequalities, they saw themselves as salaried workers and did not see differences between the working dynamics of the collective work in the settlement and the work they did for larger landowners when they were "boias-frias" (seasonal rural workers) (Scopinho, 2006).

In summary:

> Although the settlers were generally against the idea of forming a cooperative, cooperation was as a social action was very present in the daily lives of families and in small work groups that were formed as needed and just as easily dismantled. We can say that in the Sepé Tiarajú spontaneous cooperation prevailed, organically arising according to the settlers' needs. More than the ownership of a piece of land and income, the settlers hoped to obtain housing, work and physical and psychological security in Sepé Tiarajú, as well as the possibility of social and political participation.
>
> SCOPINHO, 2006

We believe that the Lula and Dilma governments at the federal level, as well as the PSDB government in São Paulo, blocked land reform in its broadest sense. However, some symbolic policies were created by the Federal Government, among which we could highlight the National Policy on Agroecology and Organic Production (PNAPO) and the "Terra Forte" ("Strong Land") Program.

These policies were formally created, but there was and is no political will for the dissemination of agroecology and cooperative experiences, as we have seen in the previous pages. For this to happen, there must be the unification of anti-capital struggles in Latin America and the overcoming of the capitalist state that sustains agribusiness.

Let's look at a news article published by the MST website: "Settlers achieve the creation of the 1st agro-industry of the Terra Forte program in Andradina – São Paulo"

The families settled near Andradina (SP) will be pioneers with the first agro-industry project financed by the Terra Forte Program in the state of São Paulo. Coapar will receive approximately R$ 12.8 million to invest in the agroindustry's construction. For Lourival Plácido de Paula, president of COAPAR, the signing of the agreement means recognition of the work carried out for years in the settlements. The Andradina region was based on beef cattle, but as the settlers and their production developed, rural workers migrated to dairy production. "Industrialization is a necessary element to add value to the raw material produced by the settlements. We hope to contribute more and more to cooperatives and to the development of Agrarian Reform", said the Minister of Agrarian Development of São Paulo, Laudemir Muller.

The Terra Forte Program, launched in 2013 by President Dilma Rousseff, aims to support agro-industrialization projects in Agrarian Reform settlements.

According to Delveck Matheus, from the MST's national leadership, the program results from years of struggle by rural social movements, and was consolidated through a joint action between the movements, the General Secretariat of the Presidency of the Republic, the Ministry of Agrarian Development, the National Supply Company (Conab) and the National Institute for Colonization and Agrarian Reform (Incra). The signing of this agreement is important to enable some old requests made by settled families.

The national president of INCRA, Carlos Guedes, recognized that "this is a project for each and every settler. Each settler who fights daily for their rights and for the recognition of their land. It is a step, not the end of the journey. Our role is to invest and support actions that prioritize the sustainable growth of the settlements. This is a sign that rural workers are on the right path".

With 46 settlements and over 4,500 families, Andradina is the only city in the country to have a Special Secretary for Agrarian Reform. For the city's mayor, Jamil Akio Ono, the investment made in the region through cooperation will benefit the entire city.

The agro-industry will be built in the industrial district of Andradina. With the effective release of the funds, the mayor expects construction to begin in early 2015. The cooperation agreement is the first of a total of 33 projects appreciated by the federal government. Of these, "23 were approved and are awaiting release of funds. The plan, which has a duration of five years, does not provide a deadline for the transfer of funds" (MST, 2015).

This blocking of agrarian reform neutralized the creation of more advanced actions for productive and distributive organization models in the settlements. While it is true that the PAA and PNAE contributed to improving the income of the settlements, it is also true that these programs gave a utilitarian and economistic character to the workers' cooperatives and associations.

To make matters worse, in the absence of the creation of general conditions of production based on associated work and agroecology, it is natural that in the few settlements created in the State of São Paulo, things operate in a logic of "every man for himself". Each family tries to produce on their piece of land in an attempt to extract the means for their immediate survival from the land. Resources for infrastructure and housing are promised but not delivered, and with that, the experience of collective life and collective work lived in the camps or in previous experiences tend to fall apart.

It is also worth noting that in the last 10 years MST-São Paulo has had enormous difficulty in mobilizing the masses to fight for land, for new ways to organize work and life. These changes stem largely from the attenuation of unemployment in the Lula-Dilma governments, the Bolsa Família Program and other social policies that have changed the scenario over the last 10 years.

6 Final Considerations: Islands/Settlements Surrounded by a Green Sea of Sugarcane and Eucalyptus

Florestan Fernandes, in his classic book "The closed circuit", sought to show in the late 1960s that the Latin American bourgeoisie had clear responses to popular struggles. Instead of incorporating and absorbing the classic demands for land, better wages and living conditions, access to public education, etc., they responded by "closing the circuit" of demands and worse than that, strangling social struggles, including by the use of torture, murder, other types of extreme violence.

In the current historical moment, the reaction of the capital installed in the State of São Paulo are clear: blocking Agrarian Reform and autocratic/violent repression of popular demonstrations, leading to a kind of "closed circuit" in short circuit, as more and more latent demands appear in society.

CHAPTER 9

The Rescue of Labour School Principles by the MST and Their Influence on Agroecology Schools

1   Introduction

The fall of the Berlin Wall led to a monumental crisis on the left. Left-wing parties began to advocate mild reforms of capitalism and the gurus of capital began to defend the "end of history". Contradictorily, when it comes to education in social movements, Brazil is one of the few countries that has been theorising anti-capitalist alternatives and rescuing the legacy of Soviet pedagogy. Researchers, leaders of social movements, professors at public universities have returned to study Soviet pedagogy, with its advances, limits and contradictions. By defending alternative, anti-capitalist forms of labour and production, the MST was practically "forced" to study alternative productive and educational experiences.

A fundamental presence in the MST's educational practices is Paulo Freire, one of the greatest popular educators in Brazil and the world. In a country like Brazil, torn apart by countless social problems, one of which was illiteracy, Freire's popular education projects had a limited impact, mainly because just as his proposals were flourishing, the 1964 coup took place. With the advance of the business-military dictatorship (1964–1985), as we have seen in previous chapters, Paulo Freire went into exile. He took part in "extension" or popular education projects in Chile, Africa and elsewhere.

Another fundamental presence in the MST's educational practices is the revival of a Latin American perspective of national sovereignty against imperialism. This is where José Marti, Che Guevara and other revolutionaries who fought for the political and economic independence of underdeveloped nations come into the MST's pedagogy. Similarly, the history of MST education is the labour school, the subject of this chapter. As we know, the Russian Revolution produced, at least in its first phase (1917–1930), one of the greatest advances in terms of criticising capitalist state education and building an alternative pedagogy, in theory and in practice. It seems that the MST has been reviving these principles in its schools and educational system. This chapter presents the foundations of the labour school. In the final part of the chapter, we try to exemplify how agroecology schools try to practice these principles, developed by the labour school.

2 The Re-release of Books from the First Phase of Soviet Pedagogy by the MST

In 2017, we commemorated the 100th anniversary of the Russian Revolution, certainly the most important historical event of the 20th century. This year we also celebrate the 150th anniversary of Karl Marx's "Capital" and the 150th anniversary of Tolstoy's "War and Peace". In 2018, we will celebrate the bicentenary of the birth of our master Karl Marx, certainly the most important intellectual of recent times. We will also celebrate the 200th anniversary of Mary Shelley's book "Frankenstein", which, incidentally, was widely used by Karl Marx.[1] It goes without saying that the pedagogy developed in the 1st phase of the Russian Revolution is being neglected in pedagogy courses and in the university in general. In a field increasingly dominated by conservative thinking, as a result of the 1964 coup, the counter-reforms of the state, and the rise of ultra-liberal thinking, Soviet pedagogy was side-lined.

We rely on the books edited by the publisher Editora Expressão Popular (The MST's publisher), which, at the initiative mainly of Professor Luiz Carlos de Freitas and Roseli Caldart, has been engaged in translating and/or revising part of the work of important Soviet intellectuals. In 2017, a collection of articles by Krupskaya – previously unpublished in Portuguese – was released. This collection was named "The construction of socialist pedagogy" and brings some of the most important essays of this magnificent militant intellectual. Some of his seminal texts had already been translated into Spanish, French and English. Amid the struggle for "redemocratization", in the early 1980s, the first edition of "Fundamentals for a Labour School", by Moisey Pistrak, was published by publisher Editora Brasiliense. In 2001, the second edition was released by publisher Editora Expressão Popular. In 2002, a book by Cecília Luedemann (2002) called "Anton Makarenko – life and work" was published. In 2005, "Pedagogical Poem", by Anton Makarenko, was published by publisher Editora 34. In 2009, the book "The Commune-School", organized by Pistrak, was published. In 2013, "Towards Polytechnism" by Viktor Shulgin. In 2014, "Essays on the Polytechnic School", Pistrak. In 2018, the 3rd edition of "Fundamentals of the Labour School" (Expressão Popular) will be released, now translated by Luiz Carlos Freitas. This is a very important book to think about and update the debate on education beyond capital.

Apparently, Brazil has proved to be a good place for the diffusion of Soviet educational thought in the first (pre-Stalinist) phase of the Russian revolution.

---

1 T.N.: This chapter was written in 2017, thus the use of the future tense.

As we are a major producer of wealth (for export) and at the same time a major producer of misery, Brazil has become a barn for alternative education proposals, especially because of the agrarian-urban issue that produces a mass of illiterate workers on a large scale. Interestingly, we have passed the 20th century without achieving literacy of the masses and, it seems, in the absence of an urgent and necessary communist revolution, we will spend the 21st century in much the same way.

3   The River that Divides the Pedagogies of Capital and the Pedagogies of Labour

It must always be remembered that there is a river that divides class society. On one side are the capitalists, their technocrats, their intellectuals and their pro-capital pedagogies, or if you prefer, those pedagogies that lie within the orbit of capital. On the other are workers and pedagogies of labour, beyond capital or outside the orbit of capital. These pedagogies, in one way or another, question the ownership of the means of production, but fundamentally the meaning of labour in societies commanded by capital.

Giving this warning is important, because to build and sustain their hegemony the proprietary classes and their technocrats try to obscure or invalidate pedagogies of labour and naturalize pro-capital pedagogies. In the oft-remembered words of Marx, the class that owns the means of production is also the producer of the ideas necessary for its reproduction. Without dominant ideas, you cannot build your hegemony. Without pedagogical production under its control, it certainly cannot reproduce itself.

The pedagogies of capital underwent significant changes from the 1st Industrial Revolution to now. Nowadays, in the face of the structural crisis of capital that does not generate jobs for everyone and destroys the material basis of life on earth, it is renewed with its "5 Pillars": "Learning to Know; Learning to Do; Learning to live with others; Learning to be; Learning entrepreneurship". There are then numerous pedagogies suitable for the phase of flexible-digital-financed accumulation, ranging from proposals for a new engaged and flexible "collaborator" who knows how to operate machines of the "digital" era, to pedagogies for "sustainable development", for "entrepreneurship", etc.

However, contradictorily, the most important pedagogy of the 21st century becomes the pedagogy of de-education, especially in countries with dependent and peripheral capitalism. Apparently, the classes that own the means of production and their managers abandoned any minimally civilizing and republican project, leaving the masses in the underworld of de-education.

As we will see later, on the other side of the river are pedagogies of labour, which can be divided into socialist and communist pedagogies. They are committed to the struggles of the workers for the emancipation of labour. We could highlight the educational experiences created by the socialists Robert Owen and Charles Fourier, German and French pedagogues. This chapter aims to address the theoretical contribution of Soviet pedagogues to the construction of communist pedagogy in the context of the Russian Revolution.[2]

4   Fundamentals of the Labour School

At the beginning of the 20th century, Russia was one of the countries where Karl Marx's contribution to a "social revolution" and an "educational revolution" was most debated. D. Lepechinsky, from a generation prior to the revolutionaries of 1917, Nadezdha Krupskaya, Anatóli Lunacharsky, Moisey Pistrak, Viktor Shulgin, Anton Makarenko and Bandlonsky are some of those who intensely debated the particularity of Russian society (and therefore of Russian education) and the socio-educational proposals of Karl Marx.

Several intellectuals characterize the period from 1900 to 1917 as a period of intense flourishing of criticism of capitalist education and the need to build an educational theory for a possible revolution in Russia. As we shall see later, many of the ideas developed in the pre-revolution phase became practical with the Russian Revolution of 1917. The Russian Revolution became, from this point of view, the most important historical milestone of the twentieth century, to the point that E. Hobsbawm (1996) determined the milestones of the twentieth century based on the Russian Revolution and its implosion in 1989–91.

It is possible to say that communist pedagogy has some foundations that we will characterize briefly in this chapter: a) it proposes a single school of labour centred on the emancipation of labour, b) focus on polytechnic education; c) self-direction, d) it follows a system of thematic complexes; e) it proposes the teaching of history from the perspective of dialectical materialism.[3]

But what are the objectives of the School of Labour? We could summarize it as the intellectual, political, polytechnic, physical and aesthetic development

---

[2] Despite some practical "rehearsals" carried out by anarchists, Charles Fourier and Robert Owen, the experimentation of the Single School of Labour on a large scale occurred only with the Russian Revolution and even then in a very limited way.
[3] We will not address the role of physical education and aesthetic education for the Soviet pedagogues of the 1st phase of the (pre-Stalinist) Revolution in this chapter.

of the workers of the communist society under construction. For them, as well as for Marx, free from the constraints of private ownership of the means of production, and assuming social control of labour with a view to the emancipation of humanity, it is possible for human beings develop through work.

4.1   The Single School of Labour

Briefly, it is possible to state that the capitalist society of the industrial era created capitalist education, mainly through the construction of state educational systems. With the First Industrial Revolution and the emergence of a "specifically capitalist mode of production", the owners of the means of production needed to create two educational systems: one for the working masses and one for the bourgeoisies and the managers-technocrats who command the production-accumulation of capital. England and France, cradles of the industrial revolution, were the first countries to create dual educational systems.

Generally speaking, class society has created an education for the sons of the bourgeoisie and middle classes distinct from the education of the workers, that is, capitalist class society has created distinct roles in production for the workers and for the capitalists and their managers. It is already possible to say that for Soviet pedagogues, this education does not allow for the intellectual, political, polytechnic, artistic and physical development of workers, condemned to work for wages.

We believe that the fundamental pillar of Soviet pedagogy is the struggle for the emancipation of labour and the role of schools in this emancipation. On the theoretical level, the need to build a society no longer based on the exploitation of labour was at stake, in view of what Marx called "self-government by freely associated producers".

Before advancing, Shulgin (2013) notes that school is one of the formative agencies of modern societies, but it is not the only one. Other training agencies are equally important: the party, the trade union, work, school, art education agencies and the family.[4] If the training-qualification complex should no longer serve to perpetuate class society, what then is the role of school in training and at work?

Regarding work, it is possible to infer that for Soviet pedagogues training takes place at work to enable emancipated labour. In this sense, work is educational (it is a formative agency). In other words, Soviet pedagogy pays special attention to the role of the school in the creation of collective workers and

---

4   In capitalism, school is one of the formative agencies but it is also accompanied by so many others. Nowadays, de-education starts very early: with cartoons, on television, in video games, on Facebook, on Whatsapp, at school, at alienated jobs and at church, among others.

their contribution to the construction of emancipated labour. For Soviet pedagogues, it was possible and necessary to prepare for collective labour from an early age, learning to cooperate to build a new country and a new society, overcoming the duality of the school system, and creating a system in which everyone can develop at work.

Pistrak (2001) divides the debate on labour into three dimensions: reproductive labour, productive labour and social labour. "Reproductive" labour is that which every human being needs to perform, unless they have a slave or a wage worker to perform it. It's also called self-service. For him, washing clothes, ironing clothes, cooking, washing dishes, cleaning the school and the home are all considered self-service. Pistrak (2001) considered this type of work fundamental to educate children to "take care of themselves" on a daily basis, as long as it did not become a boring activity, or turn them into "child slaves", in his words.[5]

In relation to productive labour, the pedagogues divided the teaching of work in school according to the specificities of the children's ages. In the first phase, children should learn self-service activities at school and practice light work activities, such as handicrafts, light joinery, etc.

Social Labour, or in the words of Shulgin (2013) "socially necessary labour"[6] is work of social interest or community interest.[7] Shulgin gives as an example of activities of social interest maintaining of public squares, taking care of schools, tending to community gardens, and doing the upkeep of roads. For him, all these activities bring about extremely complex and interesting questions in mathematical, geographical, and sociological terms that must be developed by the school, allowing theory and practice to merge.

---

5 Reproductive labour is predominantly female and hidden work. We generally do not see self-service as work, because it is considered "domestic" labour and we do not see it because it is considered female labour. Without reproductive labour, humanity would certainly not be able to survive. In the division of housework in Brazil, working women are usually the ones who take care of children, wash clothes, and go to the supermarket. Even with the entry of women into the labour market around the 1970s, this type of activity is still predominantly female, leading to the theories of double working hours. See, e.g., Krupskaya (2017), Angela Davis (2013), and Schneider (2017).
6 The category "Socially Necessary Labour" has nothing to do with the similarly-named category in political economy.
7 When reading Shulgin's (2013) work, the reader should not confuse Socially Necessary Labour with "NGOism". As always, capital transforms all community activity into "voluntary work". In the debate we had about the book in our research group, some students even transplanted the concept to the present day, as if Shulgin were an idealizer of "NGOism".

4.2 Polytechnic Schools

For Soviet pedagogues, understanding the main branches of industry in theory and practice was a challenge for revolutionary Russia. In their way of thinking, all workers have to have a general idea – it is worth insisting, in both theory and practice – of the main branches of production. According to Krupskaya (2017), children and young workers must become familiar with modern techniques.

Krupskaya (2017) questions what the content of polytechnic education consists of. She says:

> it would be a mistake to think that this content is reduced only to the acquisition of a certain number of skills or to different craft skills, as others believe, or only to the teaching of modern and higher forms of techniques. Polytechnism is a global system on the basis of which lies the study of technique in its different forms, taken in its development and in all its mediations. This includes the study of "natural technologies", as Marx called living nature, and the technology of materials, as well as the study of the means of production, their mechanisms, the study of energy. This includes the study of the geographical basis of economic relations, the impact of extraction and processing procedures on the social forms of labour, as well as their impact on the entire social order.
>
> KRUPSKAYA, 2017, p. 150–151

Krupskaya (2017) believes that

> polytechnic schools differ from professional schools by having their centre of gravity in the understanding of work processes, in the development of ability to unite theory and practice in a single form, in the ability to understand the interdependence of known phenomena, while the centre of gravity of the professional school is in the training of students in work skills.
>
> KRUPSKAYA, 2017, p. 153

According to Caldart (2013), Pistrak defended the study of the seven main industries: energy generation and sources and the extraction of the essential materials for any industry (especially metals); energy transformation; material processing; civil engineering; basic chemical industries, transport and communications; and agricultural production.

4.3   Self-direction

As is said in the educational debate, the form of the school is also a source of education. School is not only a space for unequal socialization of content according to the social classes of origin, it is also a space for preparing hierarchical social relations. In the words of Viktor Shulgin (2013), the school produces relationships. If it is true that the school teaches not only content but also relations of subordination, hierarchy, submission, so the school of labour must radically change the form of the school.

In other words, the school is not only an ideological apparatus; it is also an apparatus for reproducing social relations. From an early age the state school prepares young people who will be workers in capitalist factories, in capitalist commerce, in capitalist services, etc., for subordination. If capitalist school model forms the youth for relations of domination, command and submission, it is necessary to put a new type of school into practice, where self-organization is experienced.

Democracy was not understood as being only "outside" the school, but rather inside and outside it. The practical experience of radical democracy in school is carried out through the experimentation of self-management in the school space. More than theorizing democracy, they said, it was necessary to exercise self-direction in practice.

For them, all children and young people must go through the organizational functions of the school: learning to speak in public, deciding and respecting decisions, learning to organize an assembly, as commanders and commanded, learning to follow orders and give orders. Otherwise, there is the bureaucratization and emergence of a new social "class", the leaders, detached from the passive masses.[8]

4.4   Thematic Complexes

We believe that one of the greatest contributions of Soviet pedagogues, especially Krupskaya and Pistrak, is the theorization of complex systems. To develop the theory of complexes, these intellectuals made a radical criticism of state schools as simplifying, anti-dialectical and positivist institutions, fundamentally fragmenting the complex, dynamic and contradictory nature of reality. According to Gramsci, the fragmentation of reality is fundamental for the maintenance of hegemony.

---

8   For Soviet pedagogues, exercising self-management does not mean creating a school without teachers. For them, teachers' knowledge is fundamental to the communist school.

Apparently, the capitalist school has to produce alienation. It has to produce ignorance, to maintain the capitalist exploitation of the "majorities" by the "minorities" who own the means of production. Nowadays, its objective is to keep children and young workers alienated from the great problems of humanity, even in schools permeated by "Competence Pedagogy".

Fighting "positivist Marxism", Gyorgy Lukács, with his concept of totality, alongside Antonio Gramsci, Moisey Pistrak and Nadezdha Krupskaya, among others, defended the need to build a theory that explained reality in a dialectical, dynamic, contradictory way, which helped workers to understand the main determinants of a socio-environmental phenomenon.[9]

For us, the "heart" of the system of complexes, which unites theory with practice, is work. Unlike the abstract, merely phenomenological, idealistic, "Hegelian" connection of interdisciplinarity (currently so fashionable), it is in work that the connection between theory and practice occurs. It is by working in the concrete, "real" struggle that the relationship between theory and practice occurs, not in the "abstract interdisciplinarity of academia" (Freitas, 2009).[10] Thus, the pedagogy of the social environment, different from a unique and exclusive pedagogy of the school, considers schools as part of a system of formative agencies and the work as the "link" of materialism.

The system of thematic complexes presupposes teachers as collective workers, thinking together, planning together and evaluating together. In many experimental schools of the first phase of the revolution, teachers lived in the same space together, cultivated the land, debated texts – this certainly facilitated the organization of the thematic complexes. If this is true, the system of complexes presupposes the resumption of control of school systems by teachers, that is, to regain control of the school's productive process (what to teach,

---

9   It is worthwhile to remember the contribution of Karel Kosik (1972), in his book "Dialectics of the Concrete", to the understanding of reality in a totalizing and contradictory way.

10 Today, intellectuals outside and within the Marxist camp have contributed to this debate. Outside the Marxist camp, Edgar Morin is one of the best known. Another author who has great influence in the educational field is Jurjo Santomé. In Brazil, Ivani Fazenda's books has had a certain "success". Interestingly, in the flexible accumulation regime, there is talk of the "collaborator" who knows how to "see the whole", breaking from the little boxes in which knowledge is organized, know how to connect "specialization" with a "comprehensive analysis" – always with a view to innovation/solutions to the problems of capital accumulation. In Brazil, within the Marxist camp, Leandro Konder, José Paulo Netto, among others, and more recently Eleutério Prado have contributed to this debate. The teachers and pedagogues of the Landless Movement have given their practical and theoretical contribution to this issue as well. Schools that have partially or fully implemented the complex system already exist in Brazil, mainly in the South and under the control of the MST.

how to teach, how to evaluate, how to divide educational labour, the purposes/ meaning of the school and school methods, etc.)

We should remember that the system of complexes does not extinguish educational subjects, contrary to what common sense says. Krupskaya and Pistrak, among others, said that teachers remain fundamental to education. In the first years of what we call elementary school today, there is only one teacher, "integrating" the knowledge of different areas. In what we currently elementary school II in Brazil (6th to 9th grade), we have a body of specialized teachers, who plan how to "integrate" the different areas of knowledge from a given complex.

5   The Bureaucratization of the Russian Revolution and Its Educational Contingencies

The civil war practically decimated Russia. Eighteen countries, with numerous capitalist interests behind them, clashed with Russia after the revolution of 1917. Viktor Serge, in the book "Year One of the Revolution", describes the complete disintegration of the country: industries, railways, ports, etc. Thousands of workers died in the war, crops were destroyed, hunger was rampant, women and children were abandoned, and an especially harmful fact was that many leaders lost their lives.

It is possible to infer, after observing the books written by Pistrak, Shulgin and Krupskaya, the enormous difficulties in implementing Soviet pedagogy in the context of "war communism". With the rise of Stalinism, dualist education is reconstituted, especially from 1929 to 1931: poisoning, shooting by firing squad, defenestration and persecution of the main theorists of communist pedagogy become slogans.

To cite a few examples, Moisey Pistrak, one of the leading theorists of the school of labour, was shot in 1937. Riazanov, who discovered what would later be called Karl Marx's "Economic-Philosophical Manuscripts of 1844", was also shot. Isaac Rubin – who said that alienation did not automatically end with the end of private property – was also shot.

An extremely powerful party-state bureaucracy was created above the workers, controlling their lives and forms of labour. In the words of Mészáros (2002), the Soviet Union became a post-capitalist society and not a post-capital one, despite the genuine aspirations of its initial phase. There was, to some extent, the "expropriation of the expropriators", but the control of the fundamental activities of this society came to be in the hands of a powerful party-state bureaucracy, acting over and against workers. Notwithstanding this specific form of alienation and production of surplus-value, it must always be remembered that the USSR made significant progress in the fields of health and education.

Interestingly, Marx's dream – which is a dream shared by many of us – of a society where the state would wither to the point of being overcome and labour would emancipate itself from the yoke of capital, became in fact a great Soviet state monster, reproducing the social classes within a new guise, based on a very specific form of exploitation of labour. Again, this is why Mészáros (2000) called Soviet society post-capitalist, for capital was still in charge of society, albeit in a different way from capitalist society. Unfortunately, in the 21st century, we failed to create a new form of social control based on communal power and a radical change in the meaning of labour.

6 Experimenting with the Principles of the Labour School in the MST's Agroecology Schools

As we saw in the previous chapters, based on some examples from the agroecology schools, the MST has not mimetically transplanted the foundations of the labour school. This is no longer 1917, but the 2010s in Brazil. In any case, it can be said that the MST has rescued the principles of the labour school and adapted them to its reality.

One of them is self-organisation. From our experience as coordinators of an agroecology course at the Rosa Luxemburg School, we were able to encourage and verify, from concrete practice, the self-organisation of the school. The division into coordinating bodies, the general assembly and the rotation of roles show us that the form of the school also educates and that radical democracy can be exercised in a social movement school. More than talking about democracy, the most important thing was to practise the principles of radical and direct democracy.

The labour principle was applied in two ways. Self-service: washing, ironing, cooking, cleaning and looking after the school, etc. And agroecological "productive" labour: through cultivation practices based on the principles of agroecology. As there was no collective of teachers living in the school, it wasn't possible to plan the thematic complexes, at least in their most radical perspective. Even so, the content was designed to criticise the principles of the "green revolution" and the scientific foundations of agroecology, in theory and in practice. As it was a rural school, there were countless possibilities for carrying out agroecological work.

A detailed examination of the principles of the school of work, which I call the school of emancipated work, can be seen in the previous chapters. In the following pages we seek to summarize what education beyond capital means for us, based on the principles of the school of work and education beyond capital by István Mészáros.

7   The Urgency of an Education beyond Capital

The sociometabolism of capital can only produce barbarism. We are living in an era of barbarism, an era of destruction, characterized by a profound historical regression. Never before in the history of humanity has the precariousness of work, the multiplication of environmental disasters, hunger and misery, and functional illiteracy reached the current level. To make matters worse, capital is destroying public schools, public health, public welfare, promoting a new phase of liquidation of the few things which are still public in contemporary society.

The offensive of financial capital is producing a profound social setback of planetary breadth. Banks, insurance companies, billionaire private investors and pension funds have set the tone for capitalism since the 1970s, supported by police states, which transform life into business, public goods into merchandise and society into the market.

In Brazil, we are living in difficult and dark times. It is enough to remember the rupture of democratic legality, the return of slavery and an intense process of privatization in which public goods are sold in wholesale, cheaper and cheaper each week. At the "productive" level, we are witnessing a profound retrograde restructuring of production which has generated a new kind of primitive accumulation in the field: theft of land, extermination of the traditional communities and native peoples that remain, land grabbing, etc.

In the city, the relocation of industries in search of lower wages and less pressure from unions, outsourcing to lower costs and fragment the working class, "uberization" of workers and slave-like labour have all become part of the analyses of sociologists of labour. To further complicate this scenario, Google, Apple and Facebook, corporations that were born overnight, emerge as large companies that are at the forefront of financialised capitalism.

"Real capitalism" shows us every day that the society commanded by capital is irrational. It produces chronic unemployment and underemployment, environmental collapse, hunger and misery, social and institutionalized violence that mutually strengthen each other, overproduction crises, wars of great scope (World Wars 1 and 2) and preventive or "low intensity" wars. It reproduces and uses hierarchies between the sexes. It generates exploited labour, in its most varied forms: labour analogous to slavery, Taylorist-Fordist labour, "flexploited" labour, outsourced labour, "uberized" labour. It also leads to intense processes of migration and expulsion of young people from their countries. In turn, real socialism has shown us that the extinction of private ownership of the means of production does not necessarily lead to the birth of new communist social relations, and one of the legacies of the last century was the understanding that state or private capitalism are not options for the twenty-first century.

In the neoliberal city, at one end is the globalization of poverty and favelas, at the other mansions and gated communities. This new era, which could be called the Age of Barbarism, at the same time as it destroys the conquests of a part of the working class and exterminates social rights, also produces an intense reaction by workers, generally not publicized by the capitalist media.

One can also see that in the struggles for fundamental human rights, practical actions and causes start to emerge around (broad) self-management, associated labour, the decommodification of life, communal or collective ownership of the means of production, radical democracy and substantive equality, education beyond capital, the right to the city, food sovereignty, land for labour (and not business), agroecology and others.[11]

It is true that these struggles are still diffuse and sparse, without a common sense or direction against the sociometabolism of capital. It is also true that workers know what they do not want, but we still do not know very well where to go. There is obviously a lack of a revolutionary theory to help us on a 21st Century revolutionary journey.

---

11 Since 2007 we have carried out some courses and extension activities, of which we could highlight: 1) Extension and Specialization Courses at UNICAMP, in the field of Associated Labour, 2) Specialization Course "Public Management and Society", a product of the partnership between UFT-UNICAMP, funded by SENAES-MTE; 3) Traveling Improvement Course "Social Movements and Contemporary Crises", offered since 2014 – 7 editions have been held – a product of the partnership between the IBEC, GPOD and Unesp; 4) "Post-Middle School in Agroecology", Unesp-MST partnership (2014–2016), funded by the CNPq, 5) Mini Traveling Course "Agrarian Issues, Cooperation and Agroecology", in partnership with the MST (5 Editions), 6) Mini Traveling Course "Marx and Revolution in Marginalized Areas", held in Itaquera in 2017, in partnership with APEOESP of the East Zone, Frente Brasil Popular and Frente Povo Sem Medo, 7) Mini Course "Economics of de-education and Education beyond Capital Policy" (UNESP – 80 hours, scheduled for the second semester of 2018), in partnership with secondary students. More recently, we have the Course "Technician in Agriculture integrated to High School", in partnership with the Paula Souza Centre and the MST, financed by PRONERA-INCRA.

To conclude, regarding the school space itself, we retrieve from other writings what would be, in our view, the foundations of an Emancipated School of Labour in the 21st century:

a) The exercise of self-management at school: rotation of functions alongside collective self-management habits. This principle is based on the pedagogy of self-organization/creation of new social relations in school, that is, the form of the school forms those in contact with it. (Pistrak et al., 2009; Tragtenberg, 2006; Dal Ri; Vieitez, 2008; Freitas, 2009, Novaes; Nascimento, 2011);
b) The exercise of self-management of the educational system, including all the spheres currently alienated from teachers, employees, students and the "community". In this regard, it is enough to remember the power of the central administration in shaping the curriculum, wage policy, career plan, system supervision, evaluation, etc.;
c) Performing socially necessary labour (cleaning, cooking, gardening, etc.) (Pistrak et al. 2009; Shulgin, 2013);
d) Preparation for the struggle and insertion in present-day struggles, linking the school with its social environment (Pistrak et al. 2009; Novaes, 2012);
e) Use of a system of thematic complexes/holistic studies, to understand the sociometabolism of capital under a totalizing, dynamic and contradictory perspective, combining totality and particularity (Pistrak et al., 2009; Lukács, 2010);
f) Aesthetic education: preparation for the understanding and construction of non-market culture (Mészáros, 2006, ch. 13; Cabral, 2012; Hilsenbeck Filho, 2012; Vázquez, 2011);
g) Physical education for the development of the human body;
h) Studying history from a materialist point of view and rescueing of the Brazilian Theory of Revolution (Rodrigues, 2013; Lima Filho, Macedo, 2011);
i) Exercise of emancipated labour, that is, the school must prepare for collective and unalienated labour (Mészáros, 2002; Pistrak, 2001; Bruno, 2004; Vieitez; Dal Ri, 2001);
j) Socialization of historically accumulated knowledge, polytechnics and Sociotechnical Adequacy (Lombardi, 2011; Saviani, 2008; Ramos, 2010; Dagnino, 2008), so that workers can understand the "scientific foundations of labour". At the same time, carrying out a "balance sheet" of the technoscience of capital. As science and technology are not neutral, the realization of an "inventory" and "filter" of the

productive and destructive forces created by capital and the development of appropriate technoscience for human emancipation, with a view to the complete decommodification of society and self-government by freely associated producers (Dagnino, 2008; Novaes, 2012; Caldart, 2013; Rolo, 2012).

Conclusions

The problems facing humanity are too great to be solved within the capitalist mode of production. Based on our research into the MST, we have tried to address the environmental, agrarian and educational issues in this book. From these, we have come to some conclusions.

1-) From the point of view of social movement theory, it is possible to say that the Landless Movement is one of the most "complete" social movements today, since it combines the immediate struggle for land, agroecology and substantive equality with the broader struggle to build another mode of production, which we call ecosocialism. The Landless Movement has settlements, its own schools, partnerships with universities to train higher education students, gender, environmental and production sectors, etc. It is part of La Via Campesina and claims to be autonomous from the parties, although it has many connections with the Workers' Party (PT).

2-) From the initial struggle for land, countless "struggles within the struggle" emerged, i.e. from the main struggle for land, countless struggles to promote associated production, gender issues and agroecology, among others.

3-) Initially, agroecology was not part of the MST's project, but it co-evolved with the co-operative proposal, with a view to producing and consuming healthy food, and on a broader level, food sovereignty for the settlements and for Brazil.

4-) Brazil is one of the key countries in terms of observing the implementation of the "Green Revolution". The advance of the "Green Revolution" in Brazil has taken the struggles in the countryside to another level, distinct from the previous pattern of capital accumulation. The advance of the production of sugar cane, soya, eucalyptus, etc. has exacerbated the class struggle, discerning it from the pattern of accumulation in force until the 1950s of the old landowners. The enclosure of new lands, especially in the Amazon, since the corporate-military dictatorship, has increased conflicts over land. We call this period the new phase of "primitive" accumulation in Brazil. The particularity of the "political economy of the green revolution" in Brazil is that it was implemented during the period of the business-military dictatorship (1964–1985) and that, during the period of re-democratisation, the social movements in the countryside did not have the strength to neutralise the "industrialisation of agriculture", to the point where Brazil became the world's biggest consumer of pesticides and one of the biggest markets for the green revolution. In fact, we are a privileged space for the reproduction of the green revolution industry! What is our contradiction? We have become one of the largest exporters of commodities, but a significant portion of the population goes hungry. We have land and sunshine, but the people have nowhere to plant.

5-) The strategic defense of agroecology has led the MST, especially since the 2000s, to wage countless struggles against large transnational corporations. The defence of agroecology, from a social movement perspective, is both the result of a scientific condemnation of the disastrous consequences of the green revolution (concentration of land, increased use of pesticides, expulsion of peasants, export of primary products, subordinate industrialisation of agriculture, increase in disease, etc.) and the search for radical alternatives to production in the countryside (based on agroecological principles), i.e. the promotion of an agroecological transition in MST settlements.

6-) Agroecology schools are part of the MST's strategy, which combines struggles against the capitalist mode of production and reproduction. Realising that state schools train technicians in the paradigm of the "green revolution", the MST opted to create its own schools, autonomous from the state, where they can exercise principles of self-organisation, interdisciplinarity, the connection between emancipated work and education beyond capital, productive and reproductive work, etc. In these schools, a radical scientific critique of the green revolution is carried out and the principles and foundations of agroecology are disseminated, in close connection with the settlements (linking the school with the social environment).

7-) The MST's schools have "imbibed" from critical conceptions of education: Paulo Freire, Critical Education in Latin America and Soviet pedagogy. Many of the fundamentals developed and practised in the first period of the Russian Revolution (1917–1930) were taken up by the MST schools.

8-) The Brazilian state, as part of the apparatus of co-optation and domination, prevents the development of social movements' productive, reproductive (schools) or commercialisation experiences. The capitalist state, as part of the structure of domination, does not create the general conditions of production or public policies to develop the production and commercialisation of these experiences. Contradictorily, as part of the social struggles, the state is forced to create some public policies that favour social movements, giving rise to state policies that, although "marginal", somehow favour social movements. In turn, some civil servants (university professors, technicians in the Brazilian state bureaucracy, etc.) saw these social struggles as an opportunity to act as professionals at the service of fundamental human rights in an extremely unequal country like Brazil. This partly explains how parts of the Brazilian state are supporting these struggles by social movements to build and reproduce agroecology, associated production and agroecology schools.

9-) The MST's relationship with governments during the period of re-democratisation is not straightforward. In general, it opposed the presidents of the recent period (1985–2023), most of whom implemented neoliberal reforms. With regard to agroecology and agroecology schools, it is clear that the neoliberal governments did not support the creation and multiplication of MST schools, as part of this process, agroecology schools. One particular issue deserves to be highlighted: at the end of Fernando Henrique Cardoso's second administration (1994–2002), as part of pressure from rural social movements, Pronera (the National Programme for Education in Agrarian Reform) was created. The Lula (2003–2010) and Dilma Roussef (2011–2016) governments earmarked public funds for the dissemination of agroecology schools or courses in agroecology schools at a much greater rate than under Cardoso. However, from 2014 onwards, already in the context of preparations for the coup in Brazil, these resources became increasingly scarce. Michel Temer (2016–2018) and Jair Bolsonaro (2019–2022), in turn, ended all public policies that benefited rural and urban social movements in any way, drastically affecting the functioning of agroecology schools and the courses that took place there.

10-) The main contradiction of agroecology is that it shows in practice a new way of producing and living, even though it reproduces all the defects of capitalist forms of production. Agroecology in Brazil is, at least for now, a marginal phenomenon, but not an insignificant one. The grassroots forces that encourage agroecology are unable to stand up to the enormous power of the corporations that promote the green revolution on a daily basis and the large landowners. Of course, in a country with enormous land concentration – which results in enormous power for landowners and the transnational corporations of the "green revolution" – agroecology as part of the struggles of social movements is not welcome.

11-) Part of the middle classes, who are more aware of the health risks posed by pesticides, eat healthy food. Contradictorily, they have a very self-centred attitude and perspective. Therefore, for them, there is no problem in consuming poison-free food, opposing rural social movements and voting for Jair Bolsonaro. The people, on the other hand, because of the limited scale of agroecology and the control of land by agribusiness, go hungry or consume ultra-processed food with little nutritional value, or don't have a reasonable salary to buy agroecological products.

12-) The main limit of agroecology schools is that they can't stand up to state environmental education, which at best transmits the values and ideologies of green capitalism or some form of "sustainable development". Although these schools are qualitatively important, the hegemony of environmental education in Brazil and worldwide is pro-capital. Agribusiness foundations and the education secretariats of right-wing or extreme right-wing governments seek to spread "scientific" visions of the environmental issue from an ecocapitalist perspective. Even so, agroecology schools are fundamental experiences of the working class, and could spread if there is an international movement for the emancipation of labour, necessarily ecosocialist, led by social movements from all over the world.

Bibliography

ANDA. Principais indicadores do setor de fertilizantes. Disponível em: http://www.anda.org.br/estatistica/Principais_Indicadores_2015.pdf. Acesso em: 15 mar. 2016.

Andrade, Hayrton Francis Ximenes de. O princípio ecológico da proteção e utilização e o direito à propriedade rural no Brasil no atendimento a sua função social: uma abordagem histórico-jurídica-sistemica. 2003. Dissertação (Mestrado em Engenharia de Produção) – Universidade Federal de Santa Catarina, Santa Catarina, 2003.

Andrade, Márcia Regina; Di Pierro, Maria Clara. A construção de uma política de educação na reforma agrária. In: Andrade, Márcia Regina et al. (Orgs.). A educação na reforma agrária em perspectiva: uma avaliação do Programa Nacional de Educação na Reforma Agrária. Brasília; São Paulo: Ação Educativa, Nead, Pronera/Incra/MDA, 2004. p. 19–35.

Antunes, Ricardo. Prefácio. In: MÉSZÁROS, I. Para além do capital. São Paulo: Expressão Popular, 2002.

Antunes, Ricardo. A dialética do Trabalho: escritos de Marx e Engels. 2 ed. São Paulo: Expressão Popular, 2013.

Alonso, A. As teorias dos movimentos sociais: um balanço do debate. In: Lua Nova Revista de Cultura e Política, n.76, p.49–86, 2009.

Almeida, Benedita de; Antonio, Clésio Acilino; Zanella, José Luiz (Org.). Educação do Campo: Um projeto de formação de educadores em debate. Cascavel: Edunioeste, 2008.

Almeida, Jalcione. Prefácio. In: Altieri, Miguel. Agroecologia: bases científicas para uma agricultura sustentável. 3 ed. São Paulo: Expressão Popular, 2012.

Altieri, Miguel. Alli donde termina la retórica de la sostenibilidad comienza la agroecología. Buenos Aires: CERES, 1985.

Altieri, Miguel. Agroecologia: bases científicas para uma agricultura sustentável. 3 ed. São Paulo: Expressão Popular, 2012.

Altieri, M. *Agroecología:* bases científicas para una agricultura sustentable. Montevidéu: Nordan-Comunidad, 1999.

Altieri, Miguel; Nicholls, Clara I. Agroecologia: Resgatando a agricultura orgânica a partir de um modelo industrial de produção e distribuição. Ciência & Ambiente, n. 27, p. 141–152, 2003.

Altvater, Elmar. Existe um marxismo ecológico? In: Borón, A. (org.) A teoria marxista hoje. São Paulo: Expressão Popular-Clacso, 2007.

Antunes, Caio. A concepção de educação na obra de István Mészáros. In: Batista, E.; Novaes, H. T. (orgs.) Trabalho e reprodução social. Bauru/Marília: Canal 6/Praxis, 2013, 2ª edição.

Antunes, R. (org.) *Riqueza e Miséria do trabalho no Brasil*. São Paulo: Boitempo Editorial, 2006.
Antunes, R. Adeus ao trabalho? 15. ed. São Paulo: Cortez, 2008.
Antunes, R. Apresentação da coleção *Trabalho e Emancipação*. São Paulo: Expressão Popular, s/d.
Antunes, R. Os sentidos do trabalho: ensaios sobre a afirmação e negação do trabalho. São Paulo: Boitempo, 2002.
Antunes, R. *Palestra*. VI Colóquio Internacional Marx e Engels. Unicamp, IFCH, outubro de 2010.
Antunes, Ricardo. O caracol e sua concha – ensaios sobre a nova morfologia do trabalho. São Paulo: Boitempo editorial, 2005.
Antunes, Ricardo. Capitalismo pandêmico. São Paulo, Boitempo, 2022.
Antunes, Ricardo. Ricardo Antunes lança a obra Capitalismo Pandêmico. Jornal da Unicamp. Entrevista (07/07/2022). 2022. Obtida em: https://www.unicamp.br/unicamp/ju/noticias/2022/07/07/ricardo-antunes-lanca-obra-capitalismo-pandemico.
Azevedo, J.; Thomaz Júnior, A; Oliveira, A. M. S. A nova ofensiva do capital canavieiro e os desdobramentos para o trabalho no Pontal do Paranapanema e Alta Paulista (SP). Geografia em Atos (Online), v. 1, p. 10–17, 2006.
Balla, João Vitor Quintas *et al*. Panorama dos cursos de agroecologia no Brasil. Revista Brasileira de Agroecologia, 2014. ISSN 1980-9735.
Barreto, M. J. Territorialização das agroindústrias canavieiras na Região do Pontal do Paranapanema e os desdobramentos para o trabalho. 2012. Dissertação (Mestrado em Geografia) – Universidade Estadual Paulista Júlio de Mesquita Filho.
Barriguelli, J. C. Subsídios à história das lutas no campo em São Paulo (1870–1956). São Carlos: UFSCAR, Arquivo de História Contemporânea, 1981. 3v.
Bechara, Miguel. Extensão Agrícola. São Paulo: Secretária de Agricultura, 1954.
Benevides, Maria Vitória. Democracia e Cidadania. Revista Pólis, n. 14, São Paulo, 1994.
Benini, E. Novaes, H. T. *As lutas por uma educação para além do capital na América Latina e a criação do curso de especialização "Gestão Pública e Sociedade"*. In: Benini, E.; Sales, A. L.; Novaes, H. T.; Silva, M.R. Gestão Pública e Sociedade: balanço e resumos dos trabalhos da 3ª edição do curso de especialização. São Paulo: Outras Expressões/Cromosete, 2013.
Benini, E. Sistema orgânico do trabalho. Rio de Janeiro: Ícone, 2012.
Benini, E.; Faria, M. S.; Novaes, H. T.; Dagnino, R. (orgs.) *Gestão Pública e Economia Solidária*. São Paulo: Outras expressões, 2012.
Bensaid, D. Sobre a questão judaica. In: Marx, K. A questão judaica. São Paulo: Boitempo Editorial, 2009.
Bensaid, Daniel. Os irredutíveis. São Paulo: Boitempo editorial, 2008.
Benthien, Patrícia Faraco. Transgenia agrícola e modernidade: um estudo sobre o processo de inserção comercial de sementes transgênicas nas sociedades brasileira

e argentina a partir dos anos 1990. 272f. Tese de doutorado. Universidade de Campinas, 2010.

Bernardo, J. Economia dos conflitos sociais. São Paulo: Expressão Popular, 2006.

Bernardo, J. MST e agroecologia: uma mutação decisiva. Passa Palavra, 2012. Disponível em: <http://passapalavra.info/2012/03/97517>.

Bernardo, João. Transnacionalização do capital e fragmentação dos trabalhadores. São Paulo: Boitempo, 2002.

Bernardo, João. Democracia totalitária: teoria e prática da empresa soberana. São Paulo: Cortez, 2004.

Bernardo, João. MST e agroecologia: uma mutação decisiva. Passa Palavra, 2012. Disponível em: http://passapalavra.info/2012/03/97517. Acesso em: 30 nov. 2014.

Bertero, José Flavio. Dualismo, estagnação e dependência na América Latina: o caso do Brasil nos anos 60. Estudos de Sociologia, Araraquara, v. 3, n. 6, 1999.

Besancenot, Olivier; Lowy, Michel. Afinidades revolucionárias. São Paulo: UNESP, 2017.

Bhagavan, M. A critique of India's economic policies and strategies. Monthly Review (Vol. 39), 1987.

BIONATUR, cooperativa de produção de sementes agroecológicas. Disponível em http://www.mst.org.br/video-bionatur. Obtido em 14 de junho de 2010.

Bionatur. 2013. [online] Disponível em <http://www.youtube.com/watch?v=wGTm m9xU7LA>. Acesso em: 14 de junho de 2010.

Bonduki, Nabil. Origens da habitação social no Brasil. São Paulo: Estação Liberdade, 1999.

Borsatto, Ricardo Serra; Carmo, Maristela Simões do. A construção do discurso agroecológico no Movimento dos Trabalhadores Rurais Sem-Terra. RESR, Piracicaba, v. 51, n. 4, p. 645–660, out./dez. 2013. Impressa em fevereiro de 2014.

Brabo, Tânia Suely A. M. Sabia, Claudia P. de P. Relatório de entrevista: Simone Aparecida Resende – Escola Latino Americana de Agroecologia – ELAA. 01 out. 2014. (não foi publicado).

BRASIL. Ministério do Desenvolvimento Agrário; Instituto Nacional de Colonização e Reforma Agrária. Programa Nacional de Educação da Reforma Agrária: Manual de Operações. Brasília, 2004.

BRASIL. Ministério da Educação e do Desporto – Conselho Nacional de Educação. Ensino médio e técnico – organização curricular. PARECER Nº: CEB 009/98. Disponível em: http://portal.mec.gov.br/cne/arquivos/pdf/1998/pceb009_98.pdf. Acesso em: 30 nov. 2015.

Brunhoff, Suzane. et al. Las finanzas capitalistas: para comprender la crisis mundial. Buenos Aires: Herramienta, 2009.

Bruno, L. Estudos sobre poder político, ideologia, trabalho e educação. 2004. 512 f. Tese (Livre-Docência), FE – USP, São Paulo, 2004.

Bruno, Lucia. Estudos sobre poder político, ideologia, trabalho e educação. Livre Docência, USP, 2004.

Bunker, S. G.; Ciccantell, P. Globalization and the Race for Resources. Baltimore: Johns Hopkins University Press, 2005.

Capra, Fritjof. Sabedoria incomum. 10 ed. São Paulo: Ed. Cultrix, 1995.

Cabral, F. Arte para pensar a vida e educar os sentidos. In: Mendonça, S. G. L. et al. (Orgs.). Marx, Gramsci e Vigotski: aproximações. Araraquara, SP: Junqueira & Marin, 2012. p. 377–398.

Caldart, Roseli Salete; Cerioli, Paulo Ricardo; Kolling, Edgar Jorge. Educação do Campo: Identidade e Políticas Públicas. Por uma educação do campo. São Paulo: Anca, 2002.

Caldart, Roseli Salete. Pedagogia do Movimento Sem Terra. 3 ed. Petrópolis: Vozes, 2004.

Caldart, Roseli Salete. Elementos para a construção do Projeto Político e Pedagógico da Educação do Campo. In: ParanÁ. Secretaria de Estado da Educação. Cadernos temáticos: educação do campo. Curitiba: SEED, 2005. p. 23–34.

Caldart, Roseli Salete. Sobre a educação do campo. In: Santos, Clarice dos Santos (Org.). Educação do campo: Campo – políticas públicas – educação. Brasília, DF: INCRA; MDA, 2008.

Caldart, Roseli Salete. Educação do campo: notas para uma análise de percurso. Revista Trabalho, Educação e Saúde, v. 7, n. 1, p. 35–64, Rio de Janeiro, mar./jun. 2009.

Caldart, Roseli Salete. Desafios do Vínculo entre Trabalho e Educação na Luta e Construção da Reforma Agrária Popular. Texto apresentado como trabalho encomendado na 36ª Reunião Anual da Anped, GT Trabalho e Educação, Goiânia, 30 set. 2013.

Caldart, R. A pedagogia do Movimento Sem Terra. São Paulo: Expressão Popular, 2004.

Caldart, R. Desafios do vínculo entre trabalho e educação na luta e construção da Reforma Agrária Popular. Goiânia, 36ª Reunião Anual da Anped, GT Trabalho e Educação, 2013 (pdf).

Caldart, R. S. (org) Caminhos para a transformação escolar. São Paulo: Expressão Popular, 2009.

Caldart, Roseli S. Pedagogia do Movimento e Complexos de Estudos. In: Sapelli, M., Freitas, L. C. e Caldart, R. S. (orgs). *Caminhos para transformação da escola 3.* Organização do trabalho pedagógico nas escolas do campo: ensaios sobre complexos de estudo. São Paulo: Expressão Popular, 2015, p. 19–66.

Campos, F. A Arte da conquista – o capital internacional no desenvolvimento capitalista brasileiro (1951–1992). Marília: Lutas anticapital, 2023.

Cano, W. As raízes da concentração industrial do Estado de São Paulo. São Paulo: Unesp, 2005.

Campos, Armando. Investigación participativa: reflexiones acerca de sus fundamentos metodológicos y de sus aportes al dessarollo social. Cuadernos de agroindústria y economia rural. N. 24, p. 129–146, 1990.

Caporal, Francisco Roberto. A extensão rural e os limites à prática dos extensionistas do serviço público. 1991. Dissertação (Mestrado em Extensão Rural) – Universidade Federal de Santa Maria, Santa Maria, 1991.

Caporal, Francisco Roberto; Costabeber, José Antônio. Agroecologia e Extensão Rural: Contribuições para promoção do desenvolvimento rural sustentável. Porto Alegre, 2004. Disponível em: http://www.emater.tche.br/site/arquivos_pdf/teses/agroecologia%20e%20extensao%20rural%20contribuicoes%20para%20a%20promocao%20de%20desenvolvimento%20rural%20sustentavel.pdf.

Caporal, Francisco Roberto; Azevedo, Edisio Oliveira (Orgs.). Princípios e Perspectiva da Agroecologia. Curitiba: Instituto Federal do Paraná, 2011.

Caporal, F.; Costabeber, J. Análise Multidimensional da Sustentabilidade – uma proposta metodológica a partir da Agroecologia. Revista Agroecologia e Desenvolvimento Rural Sustentável, Porto Alegre, v. 3, n. 3, p. 70–85, 2002.

Cardoso, E. Trabalho coletivo nos assentamentos de reforma agrária. Revista da Associação Brasileira de Reforma Agrária, vol. 4, n° 3, p.140–153, set/dez, 1994.

Carone, Edgard. A República Nova (1930–1937). São Paulo: Difel, 1981.

Carter, M.; Carvalho, H. M. A luta na terra: fonte de crescimento, inovação e desafio constante ao MST. In: Carter, M. (org.) Combatendo a desigualdade social. São Paulo: Editora da UNESP, 2004.

Carrere, R.; Lovera, S. Árvores geneticamente modificadas: um passo à frente ... na direção errada. In: Lang, C. Árvores geneticamente modificadas – A ameaça definitiva para as florestas. São Paulo: Expressão Popular, 2006.

Casado, Gloria Guzmán; Sevilla Guzmán, Eduardo; Molina, Manuel Gonzalez. Introducción a la Agroecología como Desarrollo Rural Sostenible. Madrid: Ed. Mundi-Prensa, 2000.

Casado, Gloria Gusmán; Molina, Manuel González. Sobre las posibilidades del crecimiento agrario em los siglos XVIII, XIX e XX. Un estudio de caso desde la perspectiva energética. História Agrária, n. 40, p. 437–470, 2006.

Castro, Josué. Geografia da fome (o dilema brasileiro: pão ou aço). 10ª Ed. Rio de Janeiro: Antares Achiamé, 1980.

Castro, M. Aprendendo a construir um mundo novo: feminismo, agroecologia e trabalho associado na experiência da Rede Xique-Xique. Marília, mimeo (impresso), 2012.

Cerioli, P. Educação para a cooperação: experiência do curso técnico em Administração de cooperativas do MST. São Leopoldo, UNISINOS-RS, 1997 (Especialização).

Cerioli, P.; Martins, A. Caderno de Cooperação Agrícola – Sistema Cooperativista dos Assentados. n. 5. São Paulo: MST, n. 5, 1999.

Chambers, R. Rural Development: putting the last first. Essex: Longman, 1983.

Chesnais, F. La mondialisation du capital. Paris: Syros, 1995.

Chesnais, Francois. As dívidas ilegítimas – quando os bancos se apoderam das políticas públicas. 2011. http://www.ocomuneiro.com/nr13_03_francoischesnais.html.

Chesnais, F.; Serfati, C. "Ecologia e condições físicas de reprodução social: alguns fios condutores marxistas". *Revista Crítica Marxista*, São Paulo, v.1, n. 16, pp. 39-75, set. 2003.

Christoffoli, P. I. O desenvolvimento de cooperativas de produção coletiva de trabalhadores rurais no capitalismo: limites e possibilidades. Dissertação de Mestrado. Curitiba: UFPR. 2000.

Christoffoli, P. I. A cooperação agrícola nos assentamentos do MST – desafios e potencialidades. In: Benini, E.; Faria, M. S.; Novaes, H. T.; Dagnino, R. Gestão Pública e Sociedade: fundamentos e políticas públicas de economia solidária. São Paulo: Outras Expressões, 2011, vol. I. Disponível em: gestaopublicaesociedade.blogspot.com/.

Christoffoli, P. I. O desenvolvimento de cooperativas de produção coletiva de trabalhadores rurais no capitalismo: limites e possibilidades. Dissertação de Mestrado. Curitiba: UFPR. 2000.

Christofolli, Pedro Ivan. A cooperação agrícola nos assentamentos do MST: desafios e potencialidades. *In*: Rodrigues, Fabiana C; Novaes, Henrique T; Batista, Eraldo L. (Orgs.). Movimentos sociais, trabalho associado e educação para além do capital. São Paulo: Outras Expressões, 2012.

Christofolli, Pedro Ivan. Elementos introdutórios para uma história do cooperativismo e associativismo no Brasil. *In*: NOVAES, Henrique T; Mazin, Ângelo D; Santos, Lais; Questão Agrária, Cooperação e Agroecologia. São Paulo: Outras Expressões, 2015.

Ciavatta, M. Formação integrada: entre a cultura da escola e a cultura do trabalho. In: Ciavatta, M. (org.) Memória e Temporalidades do trabalho e da educação. Rio de Janeiro: Lamparina/Faperj, 2007.

Ciavatta, Maria. A formação integrada: a escola e o trabalho como lugares de memória e de identidade. Trabalho necessário [on-line], ano 3, n. 3, 2005. Disponível em: http://www.uff.br/trabalhonecessario. Acesso em: 25 out. 2015.

Cleaver, H. Leitura política do capital. Rio de Janeiro: Zahar, 1981.

COMISSÃO PASTORAL DA TERRA. Conflitos no Campo no Brasil. Goiânia: CPT, 2014.

COMISSÃO PASTORAL DA TERRA. Conflitos no Campo no Brasil. Goiânia: CPT, 2015.

COMISSÃO PASTORAL DA TERRA. Conflitos no Campo no Brasil. Goiânia: CPT, 2016.

COMISSÃO PASTORAL DA TERRA. Conflitos no Campo no Brasil. Goiânia: CPT, 2017.

Conrad, Joseph. (1902) Coração das trevas. 1902.

Costa Neto, C. P. L. Agricultura sustentável, tecnologias e sociedade. In: Costa, L. F.; Moreira, R. J.; Bruno, R. (Org.). Mundo Rural e Tempo Presente. Rio de Janeiro, 1999, v. 2, p. 299-321.

Couto, A. M. M. (2003) Greve na Cobrasma: uma história de luta e resistência. São Paulo: Annablume, 2003.

Coutrot, Thomas. Organização do trabalho e financeirização das empresas: a experiência europeia. Outubro, Campinas, n.12, 2005.

Cruz, A.C. M. da. A diferença da igualdade: a dinâmica da Economia solidária em quatro cidades do MERCOSUL. 2006. Tese (Doutorado em Economia) – Instituto de Economia, Universidade Estadual de Campinas, 2006.

Dagnino, R. Neutralidade da ciência e determinismo tecnológico. Campinas: UNICAMP, 2008.

Dagnino, R. (org.) Tecnologia Social – ferramenta para construir outra sociedade. Campinas: Instituto de Geociências-Unicamp, 2009.

Dagnino, Renato; Novaes, Henrique T. "As forças produtivas e a transição ao socialismo: contrastando as concepções de Paul Singer e István Mészáros". Revista Organizações & Democracia, Unesp, Marília, v. 7, 2007.

Dal Ri, N. M.; Vieitez, C. Educação democrática e trabalho associado no Movimento dos Trabalhadores Rurais Sem Terra e nas fábricas de autogestão. São Paulo: Ícone, Fapesp, 2008.

Dal Ri, Neusa ; Vieitez, Candido Giraldez. Gestão associada e democrática nas escolas do Movimento dos Trabalhadores Rurais Sem Terra. In: Rodrigues, Fabiana C. et al. Movimentos Sociais Trabalho Associado e Educação para além do Capital. 1 ed. v. 2. São Paulo: Outras Expressões, 2013.

Dalmagro, S. 25 de maio: vida e luta de uma escola do campo. In: ___. Alternativas de escolarização dos assentamentos e acampamentos do MST. Veranópolis, 2003.

Davis, Angela. Mulheres, raça e classe. São Paulo: Boitempo, 2013.

Del Roio, M. Os prismas de Gramsci. A fórmula política da frente única. São Paulo: Xamã, 2004.

Delgado, G. C. Capital financeiro e agricultura no Brasil. São Paulo/Campinas: Ed. Ícone/ Ed. da Unicamp, 1985.

Delgado, Guilherme. Capital financeiro e agricultura no desenvolvimento recente da economia brasileira. Tese de doutorado. Instituto de Economia, UNICAMP, 1984.

Delgado, Nelson Giordano. Política econômica, ajuste externo e agricultura. *In*: Leite, Sérgio (Org.). Políticas Públicas e Agricultura no Brasil. Porto Alegre: Editora da Universidade, UFRGS, 2001.

Deo, Anderson. Autocracia burguesa e questão agrária no Brasil. In: Pires, J. H. et. al (orgs.) Questão agrária, Cooperação e Agroecologia. Uberlândia: Navegando, 2017, volume 3.

Deo, Anderson. Uma transição à *long term*: a institucionalização da autocracia burguesa no Brasil. In: Milton Pinheiro. (Org.). Ditadura: o que resta da transição. São Paulo: Boitempo Editorial, 2014, v. 1, p. 303–330.

Dickson, D. Tecnología alternativa y políticas del cambio tecnológico. Madrid: Blume Ediciones, 1980.

DIEESE. Relatório sobre o mundo do trabalho. 2012.

Dória, F. O nordeste: 'problema nacional' para a esquerda. In: Quartim De Moraes, J.; Del Roio, M. (orgs.) História do Marxismo no Brasil – Visões do Brasil. Campinas,: Unicamp, vol. VII.

Dowbor, L. *O que acontece com o trabalho?* São Paulo, agosto de 2001. Disponível em http://ppbr.com/ld. Obtido em novembro de 10/10/2001.

Dreifuss, R. 1964: A Conquista do Estado – ação política, poder e golpe de classe. Petrópolis: Vozes, 1981.

Duarte, N. Vigotski e o "aprender a aprender": críticas às apropriações neoliberais e pós-modernas da teoria Vigotskiana. 3. ed. Campinas: Autores Associados, 2004.

Duarte, Rodrigo. Marx e a natureza em O capital. Rio de Janeiro: Loyola, 1986.

Edwards, Steve. Os "comuns" e as multidões: considerando a fotografia de cima e de baixo. Crítica Marxista, 2017, p. 9–34.

Engels, F. A situação da classe trabalhadora na Inglaterra. São Paulo: Boitempo Editorial, 2006.

Escola Milton Santos. Nossa História. Disponível em: http://atemisems.wix.com/escol amiltonsantosvc#!ems-nossa-historia/c1kjp. Acesso em: 05 fev. 2016.

Frigotto, Gaudêncio. A Formação Integrada: A Escola e o Trabalho Como Lugares de Memória e de Identidade. *In*: Frigotto, Gaudêncio; Ciavatta, Maria; Ramos, Marise. Ensino Médio integrado: concepção e contradições. 2 Ed. São Paulo: Cortez, 2010.

Faria, M. S. de. (2011) Autogestão, Cooperativa, Economia Solidária: avatares do trabalho e do capital. Florianópolis: UFSC/Em Debate.

Faria, Maurício S. Autogestão, cooperativa, economia solidária: avatares do trabalho e do capital. Florianópolis: Editora em Debate, 2011.

Faria, Maurício. S.; Novaes, Henrique T. Brazilian recovered factories: the constraints of worker control. In: Azzellini, D.; Ness, I. (Org.). Ours to máster and to own – workers control from the Comunne to the Present . 1ed.Chicago: Haymarket Books, 2010, v. 1, p. 350–372.

Fattorelli, Maria. O manejo da dívida pública. In: SicsÚ, J. (org.) (2007). Arrecadação: de onde vem? E gastos públicos: para onde vão? São Paulo: Boitempo Editorial, 2007.

Fazenda, Ivani C. A. Integração e interdisciplinaridade no ensino brasileiro: efetividade ou ideologia? São Paulo: Loyola, 1979.

Feenberg, Andrew. Transforming Technology. A Critical Theory Revisited. Oxford: Oxford University Press, 2002.

Fernandes, Bernardo. Construindo um estilo de pensamento na questão agrária: o debate paradigmático e o conhecimento geográfico. São Paulo, Unesp, Livre Docência, 2013.

Fernandes, Florestan. Problemas de Conceituação de Classes Sociais na América Latina. *In*: Zenteno, Raúl Benítez (Coord.). As Classes Sociais na América Latina. Rio de Janeiro: Paz e Terra, 1977.

Fernandes, F. Nova república? São Paulo: Zahar, 1986.

Fernandes, F. O circuito fechado. Rio de Janeiro: Globo, 2006.
Fernandes, Florestan. Anotações sobre Capitalismo Agrário e Mudança Social no Brasil. *In*:_____. Sociedade de classes e subdesenvolvimento. 5 ed. São Paulo: Global, 2008.
Fernandes, F. A revolução burguesa no Brasil. São Paulo: Contracorrente, 2020.
Ferreira, E. B.; Garcia, S. R. O. O ensino médio integrado à educação profissional: um projeto em construção nos estados do Espírito Santo e do Paraná. In: Frigotto, G. Ciavatta, M.; Ramos, M. (orgs.) Ensino Médio Integrado: concepções e contradições. São Paulo: Cortez, 2010. 2ª Ed.
Fix, Mariana. Financeirização e transformações recentes no circuito imobiliário no Brasil. Campinas, Tese de Doutorado, Instituto de Economia, 2011.
Folgado, Cleber. Os desafios da campanha contra os agrotóxicos. São Paulo: Brasil de Fato, 2013.
Fontes, V., O Brasil e o Capital imperialismo – teoria e história. Rio de Janeiro: Escola Politécnica de Saúde de São Joaquim Venâncio, Universidade Federal do Rio de Janeiro, 2010.
Fontes, Virgínia. O Brasil e o capital-imperialismo: teoria e história. Rio de Janeiro: EPSJV/UFRJ, 2010.
Fonseca, Rodrigo. Tecnologia e Democracia. *In*: Otterloo, Adalice *et al.* Tecnologias Sociais: Caminhos para a Sustentabilidade. Brasilia/DF: RTS, 2009.
Foster, John Belamy. A ecologia em Marx. Rio de Janeiro: Civilização Brasileira, 2005.
Foster, John Bellamy. A Ecologia de Marx: materialismo e natureza. trad. Maria Teresa Machado. 2 ed. Rio de Janeiro: Civilização Brasileira, 2010.
Fraga, Lais. Extensão e transferência de conhecimento: As Incubadoras Tecnológicas de Cooperativas Populares. Tese de Doutorado, Instituto de Geociências, Unicamp, 2012.
Fraga, L.; Novaes, H. T.; Dagnino, R. Educação em Ciência, Tecnologia e Sociedade para as engenharias: obstáculos e propostas. In: Dagnino, R. (org.) Estudos Sociais da Ciência e Tecnologia e Política de Ciência e Tecnologia – abordagens alternativas para uma nova América Latina. João Pessoa: EDUEPB, 2010.
Freire, Paulo, Educação como prática da liberdade. 24ªed. Rio de Janeiro: Paz e Terra. 2000.
Freire, Paulo. Extensão ou Comunicação?. 12 ed. Rio de Janeiro: Paz e Terra, 2002.
Freire, Paulo. Pedagogia do Oprimido. 37 ed. Rio de Janeiro: Paz e Terra, 2003.
Freitas. Luiz Carlos. A luta por uma pedagogia do meio: revisando o conceito. *In*: Pistrak, Moisey (Org.). A Escola-Comuna. 2 ed. São Paulo: Expressão Popular, 2013.
Freitas, L. C. A luta por uma pedagogia do meio: revisitando o conceito. *In*: Pistrak, M. M. A escola-comuna. São Paulo: Expressão Popular, 2009, p. 8–100.
Frigotto, G. Educação e crise do capitalismo real. 5ª ed. São Paulo: Cortez, 2003.
Furtado, C., A pré-revolução brasileira. Rio de Janeiro: Fundo de Cultura, 1962.

Galvão, Andreia. Marxismo e movimentos sociais. Revista Crítica Marxista, 2011, p. 107–126.

Galzerano, Luciana. Grupos empresariais e educação básica: estudo sobre a Somos Educação. Dissertação (Mestrado em Educação) – Faculdade de Educação UNICAMP. 2016.

Gennari, A. M. (1999) Réquiem ao capitalismo nacional: lei de remessas de lucros no Governo Goulart. São Paulo: Cultura Acadêmica Editora.

Gliessman, Stephen R. Agroecología: procesos ecológicos en agricultura sostenible. Turrialba: Catie, 2002.

Gomes, João Carlos Costa. As bases epistemológicas da Agroecologia. *In*: Caporal, Francisco Roberto; Azevedo, Edisio Oliveira. Princípios e perspectiva da agroecologia. Curitiba: Instituto Federal do Paraná, 2011.

Godoi, Livia M. O capital ganha asas – reestruturação produtiva no setor aeroespacial – o caso da Embraer. Marília, Unesp, Dissertação de Mestrado, 2006.

Gohn, M. da G., Teoria dos Movimentos Sociais. Paradigmas Clássicos e contemporâneos. 10ed. São Paulo: Loyola, 2012.

Gonçalves, Sérgio. Campesinato, Resistência e Emancipação: o modelo agroecológico adotado pelo MST no Estado do Paraná. 2008. Tese (Doutorado em Geografia) – UNESP, 2008.

Gonçalves, Walter Porto et. al. In: Conflitos no campo 2015. Goiânia: CPT, 2016.

Graziano Da Silva, José. Progresso técnico e relações de trabalho na agricultura. São Paulo: HUCITEC, 1981.

GRUPO de mulheres do MST invade fábrica e destrói pesquisas genéticas. http://g1.globo.com/jornal-nacional/noticia/2015/03/grupo-de-mulheres-do-mst-invade-fabrica-e-destroi-pesquisas-geneticas.html. Retirado em 10/03/2015.

Guhur, D. M. P.; TonÁ, N. *Agroecologia*. In: Caldart, R.S; Pereira, I.B; Alentejano, P; Frigotto, G. (orgs.) Dicionário de educação do campo. Rio de Janeiro, São Paulo: Escola Politécnica de Saúde Joaquim Venâncio, Expressão Popular, p. 57–66, 2012.

Guhur, Dominique. Contribuições do diálogo de saberes à educação profissional em Agroecologia no MST: desafios da educação do campo na construção do projeto popular. Dissertação (Mestrado em Educação) – Universidade Estadual de Maringá, 2010.

Guhur, Dominique. Questão ambiental e agroecologia: notas para uma abordagem materialista dialética. In: Novaes, H. T.; Mazin, A. D.; Santos, L. Questão agrária, Cooperação e Agroecologia. São Paulo: Expressão Popular, 2015, vol I.

Guhur, Dominique Michèle Perioto; Tardin, José Maria (Org.). Diálogo de Saberes, no encontro de culturas: Caderno de Ação Pedagógica. Maringá: Escola Milton Santos, 2012.

Guillerm, Alain.; Bourdet, Yvon. Autogestão: uma visão radical. Rio de Janeiro: Zahar, 1976.

Guimarães, Guilherme. Determinantes Econômicos da Evolução da Estrutura Fundiária no Brasil. 1997. Dissertação (Mestrado em Economia) – Universidade Federal Fluminense, Niterói, 1997.

Guterres, Ivani (Org.). Agroecologia Militante: Contribuições de Enio Guterres. São Paulo: Expressão Popular, 2006.

Habermas, Juger. "Autonomy and Solidarity". Entrevistas; edição e introdução de Peter Dews. London: Verso, 1986.

Hardman, F. F. (2002) Nem pátria nem patrão! Memória operária, cultura e literatura no Brasil. São Paulo: Editora da Unesp. 3ª Ed.

Harvey, David. A condição pós-moderna. 2 ed. São Paulo: Loyola, 1993.

Harvey. D. O novo imperialismo. São Paulo: Loyola, 2004.

Harvey, D. A produção capitalista do espaço. São Paulo: Annablume, 2005.

Harvey, David. Para entender O capital – livro I. São Paulo: Boitempo editorial, 2013.

Hecht, Susanna. A evolução do pensamento agroecológico. In: Altieri, Miguel. Agroecologia: bases científicas para uma agricultura sustentável. Guaíba: Agropecuária, 2002.

Henriques, F. C. Empresas Recuperadas por Trabalhadores no Brasil e na Argentina. Doutorado (Planejamento urbano e regional). UFRJ, Rio de Janeiro, 2013.

Henriques, F. C. et. al. *Empresas recuperadas pelos trabalhadores*. Rio de Janeiro: Multifoco, 2013.

Henriques, F. C. O Engenheiro na Assessoria a Empreendimentos de Autogestão. In: Schmidt, C.; Novaes, H. T. (orgs.) Economia Solidária e Transformação Social: rumo a uma sociedade para além do capital? Porto Alegre: Ed. da UFRGS, 2013.

Henriques, F.; Faria, M. S.; Novaes, H. T. "Os distintos caminhos das fábricas recuperadas no Brasil e na Argentina" In: Rodrigues, F. C.; Novaes, H. T.; Batista, E. (orgs.) Movimentos Sociais, Trabalho Associado e Educação para além do capital. São Paulo: Outras Expressões, 2012.

Hernandez, Marisela. Garcia; Araújo, Jucemary. Pronera, Ferramenta de Mudança Socioambiental nas Áreas de Assentamento. In: Sonda, C; Trauczynski, S. C. (Org.). Reforma agrária e meio ambiente: teoria e prática no estado do Paraná. Curitiba: Kairos Edições, 2010.

Heyde, J.V.D. Neem oil and neem extracts as potential insecticides for control of hemipterous rice pests. Proceedings of the Second International Neem Conference, Eschborn, Rauischholzhausen, 2003.

Hilsenbeck Filho, A. O MST no fio da navalha – dilemas, desafios e potencialidades da luta de classes. Tese de Doutorado, IFCH, Unicamp, 2013.

Hilsenbeck, Filho A. O MST e o Teatro – potencialidades pedagógicas. 2012. Obtido em: http://passapalavra.info/2012/11/66247. Retirado em 15/11/2012.

Hilsenbeck, Filho A. O MST e o teatro: potencialidades pedagógicas. 2012. Disponível em: <http://passapalavra.info/2012/11/66247>. Acesso em: 15 nov. 2012.

Hirao, F. H.; Lazarini, K.; Arantes, P. F. Metodologia de projeto arquitetônico participativo em empreendimentos habitacionais autogeridos em São Paulo –a experiência recente da assessoria técnica Usina junto aos movimentos populares de sem-teto (UMM) e sem-terra (MST). Córdoba (Argentina), 5º Seminário Latino-Americano de Ciência e Tecnologia para o Habitat, 2010.

Hirata, H. Nova divisão sexual do trabalho? São Paulo: Boitempo, 2002.

Hirata, Helena. Transferência de tecnologia de gestão: o caso dos sistemas participativos. 1994.

Hobsbawm, Eric. A era dos extremos. São Paulo: Cia das Letras, 1996.

Holyoake, G. Os vinte oito tecelões de Rochdale. Rio de Janeiro: GB, 1933.

Ianni, O. A ditadura do grande capital. Rio de Janeiro: Civilização Brasileira, 1981.

Ianni, O. O declínio do Brasil-nação. Revista de Estudos avançados, 2000, p. 51–58.

Ianni, O. Estado e capitalismo, estrutura social e industrialização no Brasil. Rio de Janeiro, Civilização Brasileira, 1965.

Ianni, Octavio. Origens agrárias do Estado brasileiro. São Paulo: Brasiliense, 2004.

Ianni, O. Estado e Planejamento econômico no Brasil. Rio de Janeiro: Ed. da UFRJ, 2009.

Iasi, M. Alienação e ideologia: a carne real das abstrações ideais. In: Del Roio, M. (org.) Marx e a dialética da sociedade civil. Marília: Oficina Universitária, 2014, p. 95–124.

Iasi, M. Educação, consciência de classe e estratégia revolucionária. Revista Universidade e Sociedade, Distrito Federal, n. 48: 122–30. Jul. 2011. (Transcrita a partir da palestra no 5º EBEM – Encontro Brasileiro de Educação e marxismo – Florianópolis, abril de 2011).

Iasi, M. Ensaios sobre consciência e emancipação. São Paulo: Expressão Popular, 2006.

IBGE. www.ibge.gov.br 2012.

Ibrahim, J. (1986) O que todo cidadão precisa saber sobre comissões de fábrica. São Paulo: Global.

Ilha Das Flores. Direção de Jorge Furtado. Porto Alegre: Casa de cinema de Porto Alegre, 1989.

Infranca, A. Trabajo, individuo, historia: el concepto de trabajo em Lukács. Buenos Aires: Herramienta, 2005.

Jinkings, Ivana. Nobile, Rodrigo. (Org.). István Mészáros e os desafios do tempo histórico. São Paulo: Boitempo Editorial, 2011.

Kapp, S. et. al. (2008). Architecture as Critical Exercise: Little Pointers Towards Alternative Practices. Obtido em: http://www.field-journal.org/uploads/file/2008 %20Volume%202%20/Architecture%20as%20Critical%20Exercise_MOM.pdf.

Karavaev, A. Brasil, Passado e Presente do Capitalismo Periférico. Trad. De K. Asryants. Moscou – URSS. Edições Progresso, 1987.

Kolling, Edgar; Nery, Irmão; Molina, Mônica Castagna. Por uma educação básica do campo. Brasília: Ed. Universidade de Brasília, 1999.

Kolling, E. J.; Vargas, M. C.; Caldart, R. MST e educação. In: MST, Boletim da Educação n. 12. – II Encontro Nacional de Educadores e Educadoras da Reforma Agrária, Dez. 2014.

Konder, Leandro. As artes da palavra: elementos para uma poética marxista. São Paulo: Boitempo, 2005.

Konder, Leandro. Marxismo e Alienação: contribuição para um estudo do conceito marxista de alienação. 2 ed. São Paulo: Expressão Popular, 2009.

Korsch, Karl. O que é socialização? Um programa de socialismo prático In: Pinheiro, M.; Martorano, L. (orgs). Teoria e prática dos conselhos operários. São Paulo: Expressão Popular, 2013, p. 141–170.

Kosik, K. Dialética do concreto. Rio de Janeiro: Paz e Terra, 1972.

Koury, A. P. (2004) Grupo Arquitetura Nova: Flávio Império, Rodrigo Lefèvre, Sérgio Ferro. São Paulo: Edusp.

Kovel, Joel.; Lowy, Michel. Manifesto ecossiocialista internacional. Revista "Capitalism, Nature, Socialism – A Journal of Socialist Ecology", 2003.

Krupskaya, Nadezhda. A construção da pedagogia socialista. São Paulo: Expressão Popular, 2017.

Kuenzer, A. Z. As mudanças no mundo do trabalho e a educação: novos desafios para a gestão. In: Ferreira, N. S. C. Gestão democrática da educação: atuais tendências, novos desafios. São Paulo: Cortez, 1998, p. 33–58.

Kuenzer, A. Z. Pedagogia da fábrica: as relações de produção e a educação do trabalhador. São Paulo: Cortez: Autores Associados, 1985.

Lamosa, R.; Loureiro, C. Agronegócio e educação ambiental: uma análise crítica. [Obtido na Internet] 2013.

Lapyda, Ives. A "financeirização" no capitalismo contemporâneo – uma discussão das teorias de François Chesnais e David Harvey. São Paulo, USP, Dissertação de Mestrado, 2011.

Leandro, J. B. Curso técnico em administração de cooperativas do MST: a concepção de educação e a influência no assentamento Fazenda Reunidas de Promissão-sp. Dissertação de Mestrado, Faculdade de Educação, Unicamp, 2003.

Leher, R. Reforma Universitária de Córdoba, noventa anos. Um Acontecimento Fundacional para a Universidade Latino-americanista. In: Sader, E; Gentili, P; Aboites, H. (compiladores). La reforma universitaria: desafíos y perspectivas noventa años después. – 1a ed. – Buenos Aires: CLACSO, 2008.

Lenin, Vladimir. Imperialismo – fase superior do capitalismo. São Paulo: Centauro, 2010.

Lima Filho, P. A. Os devoradores da ordem: exclusão social no capitalismo incompleto. In: Galeazzi, M. A. (Org.). Segurança alimentar e cidadania: a contribuição das universidades paulistas. Campinas, SP: Mercado das Letras, 1996. p. 45–77.

Lima Filho, P. A. O Projeto Universidade Popular. São Paulo, impresso (mimeo), 1999.

Lima Filho, P. A. et al. O Projeto Universidade Popular: um marxismo para o Século XXI. In: II Encontro Brasileiro de Educação e Marxismo: "Concepção e Método". Curitiba: UFPR, 2006.

Lima Filho, P. A. Carta sobre a Universidade Federal da Grande Fronteira Sul. Campinas, Impresso, outubro de 2008.

Lima Filho, Paulo A. Sobre as revoluções burguesas radicais. In: Novaes, H. T.; Dal Ri, N. (orgs.) Movimentos Sociais e Crises Contemporâneas à luz dos clássicos do materialismo crítico. Uberlândia: Navegando, 2017, volume 2.

Lima Filho, P. A. de; Macedo, R. A poeira dos mitos: revolução e contrarrevolução nos capitalismos da miséria. In: Benini, E. A.; Faria, M. S.; Novaes, H. T.; Dagnino, R. (orgs.) Gestão Pública e Sociedade: fundamentos e políticas públicas de Economia Solidária. São Paulo: Outras Expressões, 2011.

Lima Filho, P. A.; Macedo, R. A poeira dos mitos: Revolução e contrarrevolução nos capitalismos da miséria. In: Benini, É.; SardÁ De Faria, M. S.; Novaes, H. T.; Dagnino, R. Gestão pública e sociedade: fundamentos e políticas públicas de economia solidária. São Paulo: Outras Expressões, 2011. p. 150–182.

Lima Filho, Paulo A.; Novaes, Henrique T.; Macedo, Rogério F. (orgs.) Movimentos Sociais e Crises Contemporâneas à luz dos clássicos do materialismo crítico. Uberlândia: Navegando, 2017.

Leite, Sérgio (Org.). Políticas Públicas e Agricultura no Brasil. Porto Alegre. Editora da Universidade, UFRGS: 2001.

Lima, Aparecida C. Práticas educativas em agroecologia no MST/PR: processos formativos na luta pela emancipação humana. 2011. Dissertação (Mestrado em Educação) – Universidade Estadual de Maringá, Maringá, 2011.

Lima, Aparecida C. et al. Reflexão sobre a educação profissional em agroecologia no MST: desafios nos cursos técnicos do Paraná. In: Rodrigues, Fabiana C.; Novaes, Henrique T.; Batista, Eraldo L. (Orgs.). Movimentos sociais, trabalho associado e educação para além do capital. São Paulo: Outras Expressões, 2012.

Lima, A. C. Práticas educativas em agroecologia no MST/PR: processos formativos na luta pela emancipação humana. 2011. 321 f. Dissertação (Mestrado em Educação) – Universidade Estadual de Maringá, Maringá, 2011.

Lima, A. Guhur, D. Toná, N.; Noma, A. Reflexões sobre a educação profissional em agroecologia no MST: desafios dos cursos técnicos do Paraná. In: Rodrigues, F. C.; Novaes, H. T.; Batista, E. L. (orgs.) Movimentos sociais, trabalho associado e educação para além do capital. São Paulo: Outras Expressões, 2012.

Lima, A. S. A Militância Comunista e as Lutas Camponesas no Interior Paulista (1945–1958). Dissertação de mestrado. UNESP, Marília, 2009.

Lombardi, J. C. Educação e ensino na obra de Marx e Engels. Campinas, SP: Alínea, 2011.

Loureiro, Carlos. (org.) A questão ambiental no pensamento crítico. Rio de Janeiro: Quartet, 2007.

Lowy, Michel. Ecologia e Socialismo. São Paulo: Cortez, 2003.
Lowy, Michel. Ecossocialismo e planejamento democrático. Crítica Marxista, n. 28, 2009, p. 35–50.
Lowy, Michel. Crise ecológica, crise capitalista crise de civilização: a alternativa ecossocialista. Cadernos CRH 2013, p. 79–96.
Lowy, Michel. Mensagem ecológica ao camarada Marx. Cadernos Cemarx, Campinas, n. 11, 2018.
Luedemann, Cecília. Anton Makarenko – vida e obra. São Paulo: Expressão Popular, 2002.
Lukács, G. História e consciência de classe. São Paulo: Martins Fontes, 2003.
Lukács, G. Prolegômenos para uma ontologia do ser social. São Paulo: Boitempo, 2010.
Lukács, G. The process of democratization. Albany: State University of New York, 1991.
Lukács, Gyorgy. Para uma ontologia do ser social II. São Paulo. Boitempo, 2013.
Lukács, Gyorgy. Socialismo e Democratização. Rio de Janeiro: UFRJ, 2008.
Luxemburgo, Rosa. [1917] Introdução à economia política. São Paulo: Martins Fontes, 1977.
Luxemburgo, Rosa. Reforma ou Revolução? São Paulo, Ed. Expressão Popular, 1999.
Luzzi, N. O debate agroecológico no Brasil: uma construção a partir de diferentes atores sociais. 2007. Tese (Doutorado em Ciências Sociais em Desenvolvimento, Agricultura e Sociedade) – Instituto de Ciências Humanas e Sociais, Universidade Federal Rural do Rio de Janeiro, Rio de Janeiro, 2007.
Mohr, Matheus Fernando. A formação em agroecologia no MST/SC: um olhar sobre os egressos da Escola 25 de Maio de Fraiburgo – SC. 2014. Dissertação (Mestrado em Desenvolvimento rural e sociedade) – Universidade Federal de Santa Catarina, 2014.
Moreira, Rodrigo Machado. Transição Agroecológica: Conceitos, Bases Sociais e a Localidade de Botucatu/SP. 2003. Dissertação (Mestrado em Engenharia Agrícola) – Universidade Estadual de Campinas, 2003.
Morissawa, Mitsue. A história da luta pela terra e o MST. São Paulo: Expressão Popular, 2001.
Macedo, R. O governo Lula e a miséria brasileira. Tese de Doutorado. Araraquara, FCL, UNESP, 2012.
Macedo, José Rivair; Maestri, Mario. Belo Monte, uma história da guerra de Canudos. São Paulo: Expressão Popular, 2007.
Macedo, Rogério Fernandes. A destruição em massa: a tragédia da fome e da degradação dos hábitos alimentares. In: Novaes, H. T.; Santos, J.; Pires, J. H. (Orgs.) Questão agrária, cooperação e agroecologia, vol. 1. São Paulo: Outras Expressões, 2015.
Machado, Luiz Carlos Pinheiro; Machado Filho, Luiz Carlos Pinheiro. A Dialética da Agroecologia: Contribuição para um Mundo com Alimentos Sem Venenos. São Paulo: Expressão Popular, 2014.

Machado, Lucília. A politecnia nos debates pedagógicos soviéticos das décadas de 20. Teoria e Educação, n. 3, 1991, p. 151–174.

Mafort, Kelli Cristine de Oliveira. A hegemonia do agronegócio e o sentido da reforma agrária para as mulheres da Via Campesina. 2013. Dissertação (Mestrado em Ciências Sociais) – Faculdade de Ciências e Letras, Unesp – Araraquara, 2013.

Makarenko, Anton. Poema pedagógico. São Paulo: Editora 34, 2005.

Mamani, P. *Destotalización Del poder colonial/moderno – Rotación del poder y la economía otra – El caso de El Alto-Bolívia.* Mimeo, 2012.

Mamani, P. El poder anti-liberal. In: Rodrigues, F. C.; Novaes, H. T.; Batista, E. (orgs.) Movimentos Sociais, Trabalho Associado e Educação para além do capital. São Paulo: Outras Expressões, 2013, vol II.

Manfredi, S. M. Educação profissional no Brasil. São Paulo: Cortez, 2002.

Mariátegui, J. C. Sete ensaios de interpretação da realidade peruana. São Paulo: Expressão Popular, 2008.

Marini, Rui. Dialética da dependência. *In*: Traspadini, Roberta; Stedile, João Pedro (Orgs.). Rui Mauro Marini: Vida e Obra. 1 ed. São Paulo: Expressão Popular, 2005.

Marques, Luiz. Capitalismo e colapso ambiental. Campinas: Ed. Unicamp, 2015.

Martins, A. Potencialidades transformadoras dos movimentos camponeses no Brasil contemporâneo: as comunidades de resistência e superação no MST. Dissertação de Mestrado, PUC-SP, São Paulo, 2004.

Martins, José de Souza. O poder do atraso: ensaios de sociologia de história lenta. São Paulo: Hucitec, 1994.

Martins, José de Souza. O cativeiro da Terra. 9 ed. 1 reimpressão. São Paulo: Contexto, 2013.

Martins, Rodrigo. Kátia Abreu, a ministra que desmata a razão. Carta Capital, 2015. Disponível em: http://www.cartacapital.com.br/revista/832/a-ministra-desmata-a-razao-6601.html. Acesso em: 15 mar. 2016.

Martine, George; Beskow, Paulo Roberto. O modelo, os instrumentos e as transformações na estrutura de produção agrícola. *In*: Martine, George; Garcia, Ronaldo Coutinho. Os impactos sociais da modernização agrícola. São Paulo: Editora Caetés, 1987.

Marx, K. O capital. 3. ed. São Paulo: Abril Cultural, 1983, vol I.

Marx, Karl. (1867) O capital. São Paulo: Nova Cultural, 1985.

Marx, Karl. O capital: Crítica da economia política. v I. 2 Ed. São Paulo: Nova Cultural, 1985.

Marx, K. A Guerra civil na França. São Paulo: Global, 1986.

Marx, K. O Capital. 3.ed. São Paulo: Nova Cultural, 1988. v. III e IV.

Marx, K. A questão judaica. In: Manuscritos econômicos – filosóficos. Lisboa: Edições 70, 1993. p. 35–76.

Marx, K. Manuscritos econômicos-filosóficos de 1844. Lisboa: Avante, 1994.

Marx, K. O capital. São Paulo: Nova Cultural, 1996. v. I e II.

Marx, Karl. O capital. São Paulo: Nova Cultural, 1996, volume I.

Marx, K. Manuscritos Econômico-Filosóficos. São Paulo: Boitempo Editorial, 2004.

Marx, K. A miséria da filosofia. São Paulo: Expressão Popular, 2009.

Marx, K. *A questão judaica*. São Paulo: Boitempo Editorial, 2009b.

Marx, Karl. (1859). Salário, preço e lucro. In: Marx, K. Trabalho assalariado & Salário, preço e lucro. São Paulo: Expressão Popular, 2012, p. 71–133.

Marx, K. Crítica do programa de Gotha. São Paulo: Boitempo, 2012.

Marx, K. O capital. Crítica da Economia Política. Livro I. São Paulo: Boitempo, 2013.

Marx, K, Engels, Friedrich, Manifesto do Partido Comunista. São Paulo: Expressão Popular. 2008.

Marx, Karl. (1867) O capital. São Paulo: Boitempo editorial, 2010.

Marx, Karl. (1881) A luta de classes na Rússia. São Paulo: Boitempo editorial, 2013.

Mateus, D. Entrevista a Henrique Novaes, João Henrique Pires e Douglas Silva. Agudos, maio de 2015.

Mazalla Neto, W. Agroecologia e processamento de alimentos em assentamentos rurais. Campinas: Átomo e Alínea, 2013.

Mazalla Netto, Wilon. Agroecologia e Movimentos Sociais: entre o debate teórico e sua construção pelos agricultores camponeses. 280f. 2014. Tese. (Doutorado em Engenharia Agrícola), Universidade Estadual de Campinas.

Medeiros, L. S. História dos movimentos sociais no campo. Rio de Janeiro: Fase, 1989.

Medeiros, Leonilde Sérvolo de. Sem Terra, "Assentados", "Agricultores familiares": considerações sobre os conflitos sociais e as formas de organização dos trabalhadores rurais brasileiros. In: Giarracca, Norma. Una nueva ruralidad en América Latina?. Buenos Aires: Consejo Latinoamericano de Ciencias Sociales, 2001.

Medeiros, Leonilde Servolo de. Trabalhadores, sindicatos e o regime civil-militar no Brasil. In: Pinheiro, Milton (Org.). Ditadura o que resta da transição. 1 ed. São Paulo: Boitempo, 2014.

Menezes Neto, A. J. Além da terra: a dimensão sociopolítica do projeto educativo do MST. Tese de Doutorado. Faculdade de Educação, USP, 2001.

Mészáros, I. Marx: a teoria da alienação. 4. ed. Rio de Janeiro: Zahar, 1981.

Mészáros, I. A necessidade do controle social. São Paulo: Ensaio, 1987.

Mészáros, I. Filosofia, ideologia e controle social. São Paulo: Ensaio, 1993.

Mészáros, István. Para além do capital. São Paulo: Boitempo Editorial, 2002.

Mészáros, I. A educação para além do capital. São Paulo: Boitempo, 2005.

Mészáros, I. A teoria da alienação em Marx. São Paulo: Boitempo Editorial, 2006.

Mészáros, István. O desafio e o fardo do tempo histórico. São Paulo: Boitempo Editorial, 2007.

Mészáros, I. Atualidade histórica da ofensiva socialista – uma alternativa radical ao sistema parlamentar. São Paulo: Boitempo Editorial, 2008.

Mészáros, I. Filosofia, ideologia e ciência social. São Paulo: Boitempo, 2008b.

Mészáros, I. Marxismo e direitos humanos. In: Mészáros, I. Filosofia, ideologia e ciência social. Ensaios de negação e afirmação. São Paulo: Boitempo, 2004b, p.157–68.

Mészáros, István. O poder da ideologia. São Paulo: Boitempo Editorial, 2004.

Mészáros, István. A montanha que devemos conquistar. São Paulo: Boitempo, 2016.

Mészáros, István. A única economia viável. O Comuneiro: Revista Electrónica, Lisboa, n. 5, set. 2007. Disponível em: http://www.ocomuneiro.com. Acesso em: 31 maio 2021.

Mészáros, István. Para Além do Capital: rumo a uma teoria da transição. 1 ed. São Paulo: Boitempo, 2011.

Minayo, Maria Cecília de Souza. O desafio do conhecimento: Pesquisa qualitativa em Saúde. 11 ed. São Paulo: Hucitec, 2008.

Minto. Lalo Watanabe. As reformas do ensino superior no Brasil: o público e o privado em questão. Campinas: Autores Associados, 2006.

Minto, Lalo W. A Educação da "miséria": particularidade capitalista e educação superior no Brasil. São Paulo: Expressão Popular, 2015.

Molina, Monica. Educação do Campo e Pesquisa: questões para reflexão. 1. ed. Brasília: NEAD, 2000.

Molina, M.; Arelaro, L.; Wolf, S. Resumo do 2º Encontro do Residência Agrária, Brasília, agosto de 2015.

Monbeig, P. Pioneiros e fazendeiros de São Paulo. São Paulo: Hucitec, 1984.

Moniz Bandeira, L. A. O governo João Goulart: as lutas sociais no Brasil, 1961–1964. 7ª ed. Rio de Janeiro: Revan; Brasília: Ed. UnB, 2001.

Montaño, C. Terceiro Setor e Questão Social. São Paulo: Cortez, 2002.

Morais, C. S. Elementos sobre a teoria da organização no campo. Caderno de Formação, nº 11. São Paulo: MST, 1986.

Morissawa, M. A História da luta pela terra e o MST. São Paulo: Expressão Popular, 2001.

Moura, C. As rebeliões no Estado de São Paulo. In: Moura, C. Rebeliões da Senzala. São Paulo: Anita Garibaldi, 2014. 5ª edição.

Moura, Luiz H. Ciência e Agronegócio: controle capitalista da pesquisa agropecuária nacional. In: Alentejano, P.; Caldart, R. (orgs.) MST: Universidade e Pesquisa. São Paulo: Expressão Popular, 2014.

MST. A luta continua: como se organizam os assentados. Caderno de Formação n. 10. São Paulo: MST, 1986.

MST. Reforma Agrária: por um Brasil sem Latifúndio!. Textos para debate do 4 Congresso Nacional do MST. São Paulo: MST, 2000.

MST. Construindo o Caminho. São Paulo: MST, 2001.

MST. Método de Trabalho e Organização Popular. São Paulo: Setor de Formação, 2005.

MST. Regimento Interno Escola José Gomes da Silva. São Miguel do Iguaçu, 2007.

MST. Projeto Político Pedagógico: Escola José Gomes da Silva, São Miguel do Iguaçu, 2007a.

MST. Projeto Metodológico: Turma 11 do curso Técnico em Agroecologia ensino médio integrado. São Miguel do Iguaçu, 2010.

MST. Cartografia Social dos Trabalhadores Rurais Sem Terra do Assentamento Antônio Companheiro Tavares. São Miguel do Iguaçu, 2011.

MST. Proposta de Reforma Agrária Popular do MST. In: Stedile, João Pedro (Org.). A questão agrária no Brasil: debate sobre a situação e perspectiva da reforma agrária na década de 2000. 1 ed. São Paulo: Expressão Popular, 2013.

MST. Coordenação Política Pedagógica: Sistematização do seminário de avaliação dos cursos técnicos de agroecologia do Paraná. Coordenação dos setores, mar. 2013. (não foi publicado).

MST. A história da luta pela terra. 1995. Disponível em: http://www.mst.org.br/nossa-historia/94-95. Acesso em: 30 set. 2014.

MST ORGANIZAÇÃO. *Movimento dos Trabalhadores Rurais Sem Terra.* Pagina eletrônica. [São Paulo]: 02 nov. 2012. Disponível em: <http://www.mst.org.br/taxonomy/term/330>. Acesso em novembro 2012.

MST. Assentados conquistam 1º agroindústria do Terra Forte em SP. http://www.mst.org.br/2014/09/25/assentados-conquistam-1-agroindustria-do-terra-forte-em-sp.html Obtido em 20/01/2015.

MST. Construindo o Caminho. São Paulo: MST, 2001.

MST. Método de Trabalho e Organização Popular. São Paulo: Setor de Formação, 2005.

MST-PR. Escolas de Agroecologia do Paraná. São Miguel do Iguaçu, Mimeo, 2004.

Musto, M. Revisitando a concepção de alienação em Marx. In: Del Roio, M. (org.) Marx e a dialética da sociedade civil. Marília: Oficina Universitária, 2014, p. 61–94.

Nascimento, C. Do "Beco dos Sapos" aos canaviais de Catende. (Os "ciclos longos" das lutas autogestionárias). SENAES, Abril 2005. www.mte.senaes.gov.br, 2005.

Nascimento, C. Experimentação autogestionária: autogestão da pedagogia e pedagogia da autogestão. In: Batista, E. L.; Novaes, H. T. (Orgs.). Educação e reprodução social: as contradições do capital no século XXI. Bauru, SP: Canal 6; Londrina: Praxis, 2011. p. 130–166.

Nascimento, C. Experimentação autogestionária: autogestão da pedagogia e pedagogia da autogestão. In: Batista, E. L.; Novaes, H. T. (Orgs.). Educação e reprodução social: as contradições do capital no século XXI. Bauru, SP: Canal 6; Londrina: Praxis, 2011, p. 130–166.

Ness, I.; Azzellini, D. (orgs.) Ours to master and to own – Workers' Control from the Commune to the Present. New York, Haymarket books, 2011.

Netto, J. P. Introdução. In: Marx, K. Miséria da Filosofia: resposta à Filosofia da Miséria, do sr. Proudhon. São Paulo: Expressão Popular, 2009, p. 7–16.

Netto, J. P. Uma face contemporânea da barbárie. Disponível em: http://pcb.org.br/portal/docs/umafacecontemporaneadabarbarie.pdf Acesso em jan.

Netto, J. P.; Braz, M. Economia política: uma introdução crítica. São Paulo: Cortez, 2008.

Netto, José Paulo. Capitalismo e reificação. São Paulo: ICP, 2014.

Netto, José Paulo. Marx em Paris. In: Marx, Karl. Cadernos de Paris e Manuscritos econômico-filosóficos de 1844. São Paulo: Expressão Popular, 2015, p. 9–178.

Netto, José Paulo. Prefácio. In: Ziegler, Jean. Destruição em massa – geopolítica da fome. São Paulo: Cortez, 2013.

Netto, José Paulo. Uma face contemporânea da Barbárie. Rio de Janeiro, 2008.

Noble, David. America by Design. Science, Technology and the Rise of Corporate Capitalism. New York, Oxford University Press, 1977.

Norgaard, Richard B. *Development Betrayed: the end of progress and a coevolutionary revisioning of the future*. New York Routledge, 1994.

Norgaard, Richard; Sikor, Thomas. Metodologia e prática da Agroecologia. *In*: Altieri. Miguel. Agroecologia: bases científicas para uma agricultura sustentável. Guaíba: Agropecuária, 2002.

Novaes, H. T. Peasant Leagues. In: Azzellini, D.; Ness, I. International Encyclopedia of Revolution and Protest – 1500 to the Present, 2009.

Novaes, H. T. O fetiche da tecnologia – a experiência das fábricas recuperadas. São Paulo: Expressão Popular, 2010, 2ª edição.

Novaes, H. T. (Org.). O retorno do caracol à sua concha: alienação e desalienação em associações de trabalhadores. São Paulo: Expressão Popular, 2011.

Novaes, H. T. A Autogestão como Magnífica Escola: notas sobre a educação no trabalho associado. In: Batista, E. L.; Novaes, H. T. (orgs.) Educação e reprodução social: as contradições do capital no século XXI. Bauru: Canal 6/Praxis, 2011.

Novaes, H. T Reatando um fio interrompido: a relação universidade-movimentos sociais na América Latina. São Paulo: Expressão Popular-Fapesp, 2012.

Novaes, H. T. O trabalho associado como princípio educativo e a educação escolar: notas a partir das fábricas recuperadas brasileiras e argentinas. Revista HisterBr online, 2013.

Novaes, H. T. A decadência ideológica da política educacional do Estado de São Paulo – notas sobre a precarização estrutural do trabalho docente. Marília – XIV Jornada Pedagógica, 2014.

Novaes, H. T. Trabalho associado como princípio educativo e a educação escolar: notas a partir das Fábricas Recuperadas brasileiras e argentinas. Revista HistedBr on line, 2013, p. 70–88.

Novaes, Henrique Tahan. Mundo do trabalho associado e embriões de educação para além do capital. Marília: Lutas anticapital, 2018.

Novaes, Henrique; Mazin, Diogo; Santos, Lais (orgs.). Questão Agrária, Cooperação e Agroecologia. São Paulo: Outras Expressões, 2015.

Novaes, H. T.; Benini, É. Quem controla o Estado Brasileiro: o capital rentista e a dívida pública. In: Benini, E. A.; Faria, M. S.; Novaes, H. T.; Dagnino, R. (orgs.) *Gestão*

Pública e Sociedade: fundamentos e políticas públicas de Economia Solidária. São Paulo: Outras Expressões, 2012.

Novaes, H. T.; Castro, M. Em busca de uma pedagogia da produção associada. In: Benini, É.; SardÁ De Faria, M.; Novaes, H. T.; Dagnino, R. (Org.). Gestão pública e sociedade: fundamentos e políticas públicas de economia solidária. São Paulo: Outras Expressões, 2011. p. 153–188.

Novaes, Henrique Tahan. Reatando um fio interrompido – a relação universidade movimentos sociais na América Latina. São Paulo: Expressão Popular-Fapesp, 2012.

Novaes, H. T.; Christoffoli, P. I. As contradições da auto-educação no trabalho associado: reflexões a partir da experiência das fábricas recuperadas brasileiras. In: Marañon, B. (org.) Economia Solidária. Buenos Aires: Clacso, 2013c.

Novaes, H.; Mazin, A.D.; Santos, Lais (orgs.) Questão Agrária, Cooperação e Agroecologia. 1ª ed. São Paulo: Outras Expressões, 2015.

Novaes, Henrique T. Qual autogestão? São Paulo, Revista da Sociedade Brasileira de Economia Política, n. 22, maio de 2008.

Novaes, Henrique Tahan. Prefácio. In: Torres, Michelângelo. Cidadania do capital? A estratégia da intervenção social das corporações empresariais. São Paulo: Sundermann, 2017.

Nunes, Edson. Carências urbanas, reivindicações sócias e valores democráticos. Lua Nova Revista Cultura e Política, n. 17, São Paulo, 1989. Disponível em: http://www.scielo.br/scielo.php?script=sci_arttext&pid=S0102-64451989000200005. Acesso em: 31 maio 2021.

Oliveira, Ariovaldo. U. A questão da aquisição de terras por estrangeiros no Brasil– um retorno aos dossiês. Agrária, São Paulo, No. 12, 2010.

Oliveira, Ariovaldo. U. Os Agrocombustíveis e a Produção de Alimentos. In: Simonetti, M. L. (Org.). A (in) sustentabilidade do desenvolvimento: meio ambiente, agronegócio e movimentos sociais. 1ª ed. São Paulo/Marília: Cultura Acadêmica/Oficina Universitária, 2011, v. 1, p. 159–180.

Oliveira, Francisco. Elegia para uma religião. Rio de Janeiro: Paz e Terra, 2004.

Oliveira, Marcos B. de.; Lacey, H. Prefácio. *In*: Shiva, V. *Biopirataria: a Pilhagem da Natureza e do Conhecimento*. Petrópolis: Editora Vozes, 2001.

Oliveira, Marcos B. de.; Lacey, H. Prefácio. "Fórum Social busca caminhos para uma nova ciência". Reportagem de Rafael Evangelista realizada no Seminário "Tecnociência, ecologia e capitalismo" Fórum Social Mundial janeiro de 2002. Disponível em <http://www.comciencia.br/especial/fsm2/fsmII01.htm> Acesso em julho de 2002.

Oliveira, S. B. Repensando a (re)produção social do espaço – um estudo de caso da Comuna Dom Hélder Câmara – MST. Dissertação de Mestrado, UNIFESP, São Paulo, 2013.

Oliveira, Pedro Cassiano Farias de. Extensão rural e interesses patronais no Brasil: uma análise da Associação Brasileira de Crédito e Extensão Rural – ABCAR (1948–1974). 2013. Dissertação (Mestrado em História) – Universidade Federal Fluminense, 2013.

Paiva, V. Paulo Freire e o nacional-desenvolvimento. Rio de Janeiro: Civilização brasileira, 1980.

Panitch, Leo. Repensando o marxismo e o imperialismo para o século XXI. Fortaleza, Tensões Mundiais, 2014.

Pedron, Simone Tatiana. O MST e a luta por uma educação básica no campo: o Centro de Formação do Assentamento Antônio Tavares no município de São Miguel do Iguaçu/PR. 2012. Dissertação (Mestrado em História) – Universidade Estadual do Oeste do Paraná, 2012.

Peixoto, Marcus. A assistência técnica e extensão rural e a política agrícola: crise e mudança. *In*: Dantas, Bruno; Cruxên, Eliane; Santos, Fernando; Lago, Gustavo Ponce de Leon. (Org.). Constituição de 1988: o Brasil 20 anos depois – Estado e Economia em vinte anos de mudança. vol. 4. 1 ed. Brasília: ILB/Senado Federal, 2008. p. 725–761.

Pereira, V. A. Terra e poder – formação histórica de Marília. Marília: Unesp, 2005.

Petersen, Paulo. Apresentação. Dossiê Abrasco. São Paulo: Expressão Popular, 2013.

Petersen, Paulo; Tardin, José Maria; Marochi, Francisco M. Tradição (agri)cultural e inovação agroecológica: facetas complementares do desenvolvimento agrícola socialmente sustentado na região centro-sul do Paraná. AS-PTA e Fórum das Organizações dos Trabalhadores e Trabalhadoras Rurais do Centro-Sul do Paraná, 2002.

Picolotto, Everton Lazzaretti. e Piccin, Marcos Botton. Movimentos camponeses e questões ambientais: positivação da agricultura camponesa?. Revista Extensão Rural, Santa Maria, ano XV, n. 16, p. 5–36, jul./dez. 2008.

Pinassi, M. O. Da miséria ideológica à crise do capital – uma reconciliação histórica. São Paulo: Boitempo, 2009.

Pinassi, M. O. Prefácio. Rodrigues, F. C.; Novaes, H. T.; Batista, E. L. (orgs.) Movimentos sociais, trabalho associado e educação para além do capital. São Paulo: Outras Expressões, 2012. p. 7–14.

Pinassi, M. O.; Mafort, K. Os agrotóxicos e a reprodução do capital na perspectiva feminista da Via Campesina In: Rodrigues, F. C.; Novaes, H. T.; Batista, E. L. (Orgs.) Movimentos sociais, trabalho associado e educação para além do capital. São Paulo: Outras Expressões, 2012. p. 141–158.

Pinassi, Maria O.; Cruz Neto, R. G. La minería y la lógica de la producción destructiva en la Amazonia brasileña. Herramienta (Buenos Aires), v. 51, p. 121–134, 2012.

Pinassi, Maria Orlando; Mafort, Kelli. Os agrotóxicos e a reprodução do capital na perspectiva feminina da Via Campesina. *In*: Rodrigues, Fabiana C.; Novaes, Henrique T.; Batista, Eraldo L. (Orgs.). Movimentos sociais, trabalho associado e educação para além do capital. São Paulo: Outras Expressões, 2012.

Pinheiro Machado, L. C. Transição para uma agropecuária agroecológica. Porto Alegre, Mimeo, 2006.
Pinheiro Machado, L. C. Correio eletrônico enviado para Henrique Novaes. 10/10/2009.
Pinheiro Machado, Luiz C.; Pinheiro Machado Filho, L. C. A dialética da agroecologia. São Paulo: Expressão Popular, 2014.
Pinheiro, Sebastião. A máfia dos alimentos no Brasil. Porto Alegre: CREA, 2005.
Pires, João H. S. Uma análise da proposta de formação técnica para o processo de transição agroecológica na Escola "José Gomes da Silva". 2015. Dissertação (Mestrado em Educação) Faculdade de Filosofia e Ciências/Universidade Estadual Paulista "Júlio de Mesquita".
Pires, João Henrique Souza.; Novaes, Henrique Tahan. Estudo, Trabalho e Agroecologia: A Proposta Política Pedagógica dos Cursos de Agroecologia do MST no Paraná. In. *Germinal: Marxismo e Educação em Debate*, Salvador, v. 8, n. 2, p. 110–124, dez. 2016.
Pistrak, M. Fundamentos da Escola do Trabalho. São Paulo: Expressão Popular, 2001.
Pistrak, Moisey M. Fundamentos da escola do Trabalho. 4 ed. São Paulo: Expressão Popular, 2005.
Pistrak, M. M. (Org.). A escola-comuna. São Paulo: Expressão Popular, 2010.
Ploeg, Jan. Camponeses e Impérios Alimentares: lutas por autonomia e sustentabilidade na era da globalização. Porto Alegre: UFRGS, 2008.
Porto-Gonçalves, Carlos Walter; Cuin, D. P. Os Cerrados e os Fronts do Agronegócio no Brasil. Conflitos no Campo Brasil, v. 2017, p. 74–85, 2017.
Porto-Gonçalves, Carlos Walter; Cuin, D. P.; Leal, L. T.; Silva, M. N. Bye bye Brasil, aqui estamos: a reinvenção da questão agrária no Brasil. Conflitos no Campo Brasil, v. 1, p. 86–98, 2016.
PPP. Projeto Político-Pedagógico Escola José Gomes da Silva. São Miguel do Iguaçu, Mimeo, 2010.
Prado Jr., Caio. A revolução brasileira. São Paulo: Brasiliense, 2002.
Pretto, José. Miguel. Amplitude e restrições ao acesso de PRONAF investimento no Rio Grande do Sul – Um estudo das três operações de financiamento envolvendo Cooperativas de Crédito Rural, Cooperativas de Produção Agropecuária e o Banco Regional de Desenvolvimento do Extremo Sul. 2005. Dissertação (Mestrado) – FCE – URGS, 2005.
Primavesi, Ana. Agricultura sustentável. São Paulo: Nobel, 1986.
Puigrós, A.; Gagliano, R. La fábrica del conocimiento – los saberes socialmente productivos en América Latina. Rosario: Homo Sapiens, 2004.
Quijano, A. Sistemas alternativos de produção? In: Santos, B.S. (org.) Produzir para viver, os caminhos da produção não capitalista. Rio de Janeiro: Civilização Brasileira, p. 300–345.

Rago, L. M. O controle da fábrica: os anarquistas e a autogestão. In: Rago, L. M. Do cabaré ao lar: a utopia da cidade disciplinar: Brasil 1890–1930. Rio de Janeiro: Paz e Terra, 1985.

Ramos, M. Trabalho, educação e correntes pedagógicas no Brasil. Rio de Janeiro: EPSJV/UFRJ, 2010.

Ramos, M. Trabalho, educação e correntes pedagógicas no Brasil: um estudo a partir da formação dos trabalhadores técnicos da saúde. Rio de Janeiro: UFRJ, 2010.

REDE XIQUE XIQUE. Disponível em <http://redexiquexique.blogspot.com.br/p/instituicao.html>. Acesso em 10/04/2013.

Rego, Thelmely Torres. Formação em agroecologia. Programa do Contestado da AS-PTA. 313f. Tese (Doutorado em Educação), UFSC, 2016.

Sader, Eder. Quando novos personagens entram em cena. Rio de Janeiro: Paz e Terra, 1988.

Ribas, A. D. MST, Cooperativismo e território: dinâmica e contradições. In: Thomaz Jr., A. (Org.). Geografia e Trabalho no Século XXI. Presidente Prudente: Centelha, 2004, v. 1, p. 8–40.

Ribeiro, D. Confissões. São Paulo: Companhia das Letras, 1997.

Ribeiro, L. *Via Campesina, Soberania Alimentar e Agroecologia*. XIV Jornada do Trabalho, Ourinhos, 2013.

Rocha, G. (2004) Revolução do cinema novo. São Paulo: Cosac Naify.

Rodrigues, Fabiana de Cássia. "MST Formação política e reforma agrária nos anos de 1980". 2013. Tese (Doutorado em Educação) – Universidade Estadual de Campinas, 2013.

Rodrigues, F. C. Educação política dos trabalhadores: a centralidade da exploração do trabalho na análise da questão agrária no Brasil – a contribuição de Octávio Ianni. In: Rodrigues, F. C.; Novaes, H. T.; Batista, E. (orgs.) Movimentos Sociais, Trabalho Associado e Educação para além do capital. São Paulo: Outras Expressões, 2013, vol. II.

Rodrigues, Fabiana de Cássia. A questão agrária no Brasil – as contribuições de Caio Prado Júnior, Florestan Fernandes e Octavio Ianni. In: Novaes, Henrique T.; Mazin, Ângelo D.; Santos, Lais (Orgs.). Questão Agrária, Cooperação e Agroecologia. 1 ed. São Paulo: Outras Expressões, 2015.

Rodrigues, F. C. MST – Formação Política e Reforma Agrária nos anos de 1980. Tese (Doutorado em Educação). UNICAMP, Campinas, 2012.

Rodrigues, F. C.; Novaes, H. T.; Batista, E. L. (orgs.) Movimentos sociais, trabalho associado e educação para além do capital. São Paulo: Outras Expressões, 2012. v. I.

Rodrigues, F. C.; Novaes, H. T.; Batista, E. L. (Orgs.). Movimentos sociais, trabalho associado e educação para além do capital. São Paulo: Outras Expressões, 2012. v. I.

Rodrigues, F. C.; Novaes, H. T.; Batista, E. L. (Orgs.). Movimentos sociais, trabalho associado e educação para além do capital. São Paulo: Outras Expressões, 2014. v. III.

Roger, E. Diffusion of inovations. New York, Macmillam, 1987.
Rolo, M. Ocupando os latifúndios do saber: subsídios para o ensino da ciência na perspectiva politécnica da educação. 2012. 382f. Tese (Doutorado em Educação). Rio de Janeiro, UERJ, 2012.
Rosar, M. F. Centros de Ensino Médio Integrados na região da Baixada Maranhense: pontos de desenvolvimento territorial? In: Lombardi, J. C.; Saviani, D. (orgs.) História, Educação e Transformação: tendências e perspectivas para a educação pública no Brasil. Campinas: Autores Associados, 2011.
Roy, Arundhati. Power Politics. South End Press, 2001.
Ruggeri, A. (org.) La economía de los trabajadores – Autogestión, cooperativas y empresas recuperadas en tiempos de crisis global. Buenos Aires: Continente/ Peña Lillo, no prelo.
Ruggeri, Andrés. Que son las empresas recuperadas por sus trabajadores? Buenos Aires: Continente, 2014.
Ruschel, V. B. Pedagogia da organização coletiva: a cooperativa dos estudantes da Escola Agrícola 25 de Maio. In: Vendramini, C. R. (Org.). Educação em movimento na luta pela terra. Florianópolis, SC: UFSC-CED, 2002.
Sachs, I. Espaços, tempos e estratégias de desenvolvimento. São Paulo: Vértice, 1986.
Sachs, I. *Estratégias de transição para o século 21*. São Paulo, Studio Nobel-Fundap, 1993.
Sader, Eder. Quando novos personagens entram em cena. Rio de Janeiro: Paz e Terra, 1988.
Sampaio Jr. P. A. *Entre a Nação e a Barbárie*. Rio de Janeiro: Vozes, 1996.
Sampaio Jr., P. Notas críticas sobre a atualidade e os desafios da questão agrária. In: Stedile, J. P. (org.) Debates sobre a situação e perspectivas da reforma agrária na década de 2000. São Paulo: Expressão Popular, 2013, p. 189–240.
Sampaio Jr., P. Prefácio. In: Campos, F. A Arte da conquista – o capital internacional no desenvolvimento capitalista brasileiro (1951–1992). Marília: Lutas anticapital, 2023.
Sampaio Jr., Plínio de Arruda. Notas críticas sobre a atualidade e os desafios da questão agrária. *In*: Stedile, João Pedro (Org.). A questão agrária no Brasil: debate sobre a situação e perspectiva da reforma agrária na década de 2000. 1 ed. São Paulo: Expressão Popular, 2013.
Sanfelice, J. L. A política educacional do Estado de São Paulo: apontamentos. Nuances (UNESP Presidente Prudente), v. 18, p. 145–160, 2010.
Sanfelice, J. L., Movimento Estudantil. A UNE na resistência ao golpe de 1964. Campinas: Alínea, 2008.
Santomé, Jurjo T. A Globalização e a Interdisciplinaridade: O Currículo Integrado. Porto Alegre: Artmed, 1998.
Santos, L. Gestão democrática e participação na educação profissional agroecológica do MST (PR): limites e possibilidades de uma educação emancipatória. 150f. 2015. Dissertação (Mestrado em Educação), UNESP, 2015.

Santos, Selma Aparecida dos. A trajetória do Assentamento Reunidas: o que mudou? 133f. (Dissertação de Mestrado). Instituto de Economia, Unicamp, 2007.

Santos, Selma. História da Coopava. In: Novaes, H. T.; Santos, J.; Pires, J. H. (Orgs.) Questão agrária, cooperação e agroecologia, vol 11. São Paulo: Outras Expressões, 2015.

Santos, Theotônio dos. A Teoria da Dependência: Balanço e Perspectivas. Rio de Janeiro: Civilização Brasileira, 2000.

Saviani, D. O legado educacional do "longo século XX" brasileiro. In: Saviani, D. et al. (org.) O legado educacional do século XX. 2ª ed. Campinas, SP: Autores Associados, 2006, p. 9–57.

Saviani, D. Escola e Democracia. (Edição Comemorativa). Campinas: Autores Associados, 2008. 20ª Edição.

Saviani, D. História das Ideias Pedagógicas no Brasil. Campinas: Autores Associados, 2007.

Saviani, D. O choque teórico da Politecnia. Trab. educ. saúde [online]. 2003, vol.1, n.1, p. 131–152.

Saviani, Dermeval. Trabalho e Educação: Fundamentos histórico-ontológicos da relação trabalho e educação. Revista Brasileira de Educação, Rio de Janeiro, Anped, v. 12, n. 34, Scielo, jan./abr. 2007. ISSN 1413-2478. versão impressa.

Saviani, D. Pedagogia histórico-crítica: primeiras aproximações. Campinas, SP: Autores Associados, 2008.

Schiochet, Valmor. Da Democracia à autogestão: Economia Solidária no Brasil. In: Benini *et al*. Gestão Pública e Sociedade: Fundamentos e Políticas públicas de Economia Solidária. 1 ed. v. 2. São Paulo: Outras Expressões, 2012.

Schor, J. B. Nascidos para comprar – uma leitura essencial para orientarmos nossas crianças na era do consumismo. São Paulo: Gente, 2009.

Schwarz, R. Cultura e Política – 1964–1969. Rio de Janeiro: Paz e Terra, 2007.

Scopinho, R. Controle do trabalho e condições de vida em assentamentos rurais: possibilidades e limites da cooperação autogestionária. Controle do Trabalho e Condições de Vida em Assentamentos Rurais: Possibilidades e Limites da Cooperação Autogestioária. In: VII Congreso Latino Americano de Sociología Rural, 2006, Quito-Equador. VII Congreso Latino Americano de Sociología Rural. Quito-Equador: Alasru, 2006.

Serafim, Milena. Agricultura Familiar no Brasil: uma "Análise de Política" de Políticas e Instituições. Campinas, Unicamp, Tese de doutorado, 2012.

Sevá Filho, A. Populações e Territórios espoliados pela ampliação recente da infraestrutura industrial capitalista: focos de luta política e ideológica na América do Sul. Marília: Lutas anticapital, 2019.

Sevá Filho, Oswaldo. Populações e territórios espoliados pela ampliação recente da infraestrutura industrial capitalista – focos de resistência política e ideológica na América do Sul. Marília: Lutas anticapital, 2019.

Sevilla Guzman, Eduardo; Molina, Manuel. Sobre a evolução do conceito de campesinato. São Paulo: Expressão Popular, 2011.
Sevilla Guzmán, Eduardo. La Agroecología como Estrategia Metodológica de Transformación Social. España: Instituto de Sociología y Estudios Campesinos de la Universidad de Córdoba, [s/d].
Sevilla Guzmán, Eduardo. Uma estratégia de sustentabilidade a partir da agroecologia. Agroecologia e Desenvolvimento Rural Sustentável, Porto Alegre, v. 2, n. 1, p. 35 – 45, 2001.
Sevilla Guzmán, Eduardo ; Ottmann, Graciela. Las dimensiones de la agroecología. *In*: INSTITUTO DE SOCIOLOGÍA Y ESTUDIOS CAMPESINOS. Manual de olivicultura ecologíca. Córdoba: Universidad del Córdoba, 2004. p. 11–26.
Sevilla Guzmán, Eduardo. Sobre los orígenes de la agroecología en el pensamiento marxista y libertario. La Paz: Plural Editores, 2011.
Sevilla Guzmán, Eduardo; Woodgate, Graham. Agroecología: Fundamentos del pensamiento social agrario y teoría sociológica. AgroecologAgroecología: Fundamentos del pensamiento social agrario y teorAgroecología sociolía, n. 8, p. 27–34, 2013.
Sevilla Guzmán, Eduardo ; Molina, Manuel. Sobre a Evolução do Conceito de Campesinato. 2 ed. São Paulo: Expressão Popular, 2013.
Shelley, M. Frankenstein. Porto Alegre: L&PM, 2000.
Shiva, V. Biopirataria: a Pilhagem da Natureza e do Conhecimento. Petrópolis: Editora Vozes, 2001.
Shulgin, Viktor. Rumo ao politecnismo. São Paulo: Expressão Popular, 2013.
Siliprandi, Emma. Mulheres e Agroecologia: a construção de novos sujeitos políticos na agricultura familiar. 2009. 292f. Tese (Doutorado) – Desenvolvimento Sustentável, UnB, 2009.
Silva, José Graziano da. A modernização dolorosa: estrutura agrária, fronteira agrícola e trabalhadores rurais no Brasil. Rio de Janeiro: Zahar editores, 1981.
Silva, Lígia Osório. As leis agrárias e o latifúndio improdutivo. Revista São Paulo em Perspectiva, v. 1, n. 2, abr./ jun. 1997.
Silva, N. F.; Lima Filho, P. A. A sociedade comunista na visão de Marx e Engels. In: Novaes, H. T.; Mazin, A. D.; Santos, L. (orgs.) Questão agrária, cooperação e agroecologia. São Paulo: Expressão Popular, 2015.
Singer, Paul. A recente ressurreição da Economia Solidária. In: Santos, B.S. (org.) Produzir para viver, os caminhos da produção não capitalista. Rio de Janeiro: Civilização Brasileira, 2002, p. 81–129.
Souza, Rafael Bellan. A mística no MST: mediação da praxis formadora de sujeitos históricos. 320f. Doutorado em Sociologia. UNESP – Araraquara, 2012.
Souza, S. M. R. A emergência do discurso do agronegócio e a expansão da atividade canavieira: estratégias discursivas para a ação do capital no campo. 2011. Tese (Doutorado em Geografia) – Universidade Estadual Paulista Júlio de Mesquita Filho.

Stédile, João Pedro. A questão agrária no Brasil. São Paulo: Atual (Espaço e debate), 1997.
Stédile, J. P., Fernandes, B. M. Brava Gente: a trajetória do MST e a luta pela reforma agrária no Brasil. São Paulo: Perseu Abramo, 1999.
Storch, Sergio. Discussão da Participação dos Trabalhadores na Empresa. In: Fleury, M.T. e Fischer, R.M (Org.) Processo e Relações de Trabalho no Brasil. São Paulo: Ed. Atlas, 1985.
Svampa, M., Pereyra, S. Entre la ruta y el barrio: la experiência de las organizaciones piqueteras. 1º ed. Buenos Aires: Biblos, 2003.
Tait, Marcia. Elas dizem não! Mulheres camponesas e resistência ao cultivo de transgênicos no Brasil e Argentina. 220f. 2014. Tese. (Doutorado em Política Científica e Tecnológica), Universidade Estadual de Campinas.
Tardin, J. M. Brasil não tem política para agroecologia [Entrevista]. Obtido em: http://www.mst.org.br/node/1863. 10/03/2013.
Tardin, José Maria. Camponesas e Camponeses em Movimento Construindo o Sustento da Vida e a Transformação da Sociedade. *In*: CONGRESSO BRASILEIRO DE AGROECOLOGIA, 6., CONGRESSO LATINOAMERICANO DE AGROECOLOGIA (Agricultura familiar e camponesa: experiências passadas e presentes construindo um futuro sustentável), 2., 2009, Curitiba. Anais [...]. Curitiba: ABA, SOCLA, Governo do Paraná, 2009, p. 213–217 (1 CD-ROM).
Tarrow, S., O poder em movimento. Movimentos Sociais e Confronto Político. São Paulo: Vozes, 2009.
Tavares, J. C. Universalidade e singularidade do espaço transitório: um estudo a partir de quebradeiras de coco babaçu/MIQCB e trabalhadores rurais sem terra/MST no Maranhão.
Tavares, Maria da Conceição. Império, território e dinheiro. In: Fiori, José Luis (Org.). Estados e moedas no desenvolvimento das nações. Petrópolis, RJ: Vozes, 1999.
Teixeira, Elizabeth. Reflexões Sobre o Paradigma Holístico e Holismo e Saúde. Rev. Esc. Enf. USP, v. 30, n. 2, p. 286–290, 1996.
Theis, I. M.; Meneghel, S. M. *Universidade, desenvolvimento e meio ambiente*. In: Wulf, C.; Bryan, N. P. (Org.). Desarrollo sustenible: conceptos y ejemplos de buenas prácticas en Europa y América Latina. Münster: Waxmann Verlag GmbH, 2006, v. 22, p. 85–97.
Thiollent, Michel. Metodologia da pesquisa-ação. São Paulo: Cortez, 2000.
Tiriba, L. Pedagogia(s) da produção associada. Ijuí: Ed. da Unijuí, 2001.
Tiriba, L; Fischer, M.C.B. Saberes do trabalho associado. In: Cattani, A. D.; Laville, J. L.; Gaiger, L. I.; Hespanha, P. Dicionário Internacional da Outra Economia. São Paulo/Coimbra, Almedina Brasil Ltda/Edições Almedina S.A., 2009, p. 293–298.
Tolentino, C. O farmer contra o Jeca – O projeto de revisão agrária do Governo Carvalho Pinto. Marília: Oficina Universitária, 2011.

Toná, Nilciney. A Pesquisa nos Cursos de Agroecologia e nas Escolas e Centros de Formação dos Movimentos Sociais do Campo no Paraná. *In*: SEMINÁRIO NACIONAL O MST E A PESQUISA (ITERRA), 2. Cadernos do Iterra, ano VII, n. 14, nov. 2007.

Toná, Nilciney; Guhur, Dominique Michèle Perioto. O diálogo de saberes na promoção da agroecologia na base dos movimentos sociais populares. Rev. Bras. de Agroecologia, v. 4, n. 2, p. 3322–3325, nov. 2009.

Toná, Nilciney; Guhur, Dominique Michèle Perioto. Agroecologia. In: Caldart, Roseli Salete; Pereira, Isabel Brasil; Alentejano, Paulo; Frigotto, Gaudêncio. Dicionário da Educação do Campo. RJ/SP: Escola Politécnica de Saúde Joaquim Venâncio, Expressão Popular, 2012. p. 59–67.

Toná, Nilciney. Elementos de Balanço do Processo de construção da Agroecologia no MST – PR. In: Queiroz, J. G; TonÁ, N. Nossa História: Agroecologia e as Escolas de Formação do Paraná. Curitiba: Acap., 2011.

Tragtenberg, Maurício. Sobre educação, política e sindicalismo. 2ª Ed. São Paulo: Editora da Unesp, 2006.

Tristão, Martha. A educação ambiental na formação de professores: rede de saberes. São Paulo: Annablume, 2004.

Torres, Michelangelo. Cidadania do capital? A estratégia da intervenção social das corporações empresariais. São Paulo: Sundermann, 2017.

Valadão, Adriano da Costa. Educação do Campo e Ensino Superior Tecnológico: a experiência da Escola Latino Americana de Agroecologia. In: Souza, Maria Antonia (Org.). Prática educativas do/no campo. v. 1. Ponta Grossa: UEPG, 2011. p. 41–66.

Vasconcellos, Bruna. Mulheres rurais, trabalho associado e agroecologia. In: Novaes, Henrique; Mazin, Angelo; Santos, Lais. (Org.). Questão Agrária, Cooperação e Agroecologia. 1ed. São Paulo: Outras expressões, 2015, v. 1, p. 341–370.

Vázquez, Adolfo Sánchez. Filosofia da práxis. 2 ed. São Paulo: Expressão Popular, 2011.

Vazzoler. M. R. Cooperativismo em assentamentos de reforma agrária: a sobrevivência de cooperativas do MST no contexto capitalista. 2004. Tese (Doutorado em Engenharia de Produção) – Programa de Pós-graduação em Engenharia de Produção, Universidade Federal de São Carlos, São Carlos, 2004.

Verdério, Alex. A materialidade da Educação do Campo e sua incidência nos processos formativos que a sustentam: uma análise acerca do curso de Pedagogia da Terra na UNIOESTE. 2011. Dissertação (Mestrado em Educação) – Universidade Estadual do Oeste do Paraná, 2011.

Vendramini, C. R.; Machado, I. F. (Orgs.). Escola e movimento social: experiências em curso no campo brasileiro – reimpressão. 2. ed. São Paulo: Expressão Popular, 2013.

Venturelli, R. M. Terra e poder – as disputas entre agronegócio e resistência camponesa no sudoeste paulista – uma abordagem do uso das terras públicas. Dissertação de mestrado. São Paulo, USP, 2013.

Vieira, Carlos Cordovano. Passado colonial e reversão no Brasil contemporâneo. In: Novaes, H. T.; Macedo, R. F.; Castro, F. (orgs.) Introdução à crítica da economia política. Marília: Lutas anticapital, 2019.

Vieitez, Candido.; Dal Ri, Neusa M. Trabalho associado. Rio de Janeiro: DP&A, 2001.

Viola, S. Palestra no Seminário de Direitos Humanos. Marília, Unesp, out. 2012.

Xavier, Maria Elizabete. Para um exame das relações históricas entre capitalismo e escola no Brasil: algumas considerações teórico-metodológicas. In: Zanardini, Isaura Monica Souza; Orso, Paulino José. Estado, Educação e Sociedade Capitalista. Cascavel: Edunioeste, 2008.

Xavier, Maria Elizabete Sampaio Prado. Capitalismo e escola no Brasil: a constituição do liberalismo em ideologia educacional e as reformas do ensino (1931 1961). Campinas: Papirus, 1990.

Wallerstein, Immanuel. Uma política de esquerda para o século XXI? ou teoria e práxis novamente. In: Loureiro, I.; Leite, J.C.; Cevasco, M. (orgs.) O espírito de Porto Alegre. São Paulo: Paz e Terra, 2002.

Wallerstein, Immanuel. World-Systems Analysis: an introduction. Duke University Press. 2004.

Welch, C. A. A semente foi plantada – as raízes paulistas do movimento sindical camponês no Brasil, 1924–1964. São Paulo: Expressão Popular, 2010.

Wirth, I. G.; Fraga, L.; Novaes, H. T. Educação, Trabalho e Autogestão: limites e possibilidades da Economia Solidária. In: Batista, E. L.; Novaes, H. T. (orgs.) Educação e reprodução social: as contradições do capital no século XXI. Bauru: Canal 6/ Praxis, 2011.

Ziegler, Jean. Destruição em massa – geopolítica da fome. São Paulo: Cortez, 2013.

Films

A classe operária vai ao paraíso. Direção: Elio Petri. Roma: 1976.
A corporação. 2002. Direção de Jeniffer Abbott e Mark Achbar. 2002. Canadá. Dvd.
A terceira morte de Joaquim Bolivar. Diretor: Flávio Candido (1999).
ABC da Greve. Diretor: Leon Hirszman (1979).
Acervo (Tempo Glauber) http://www.tempoglauber.com.br/english/.
Barravento. Diretor: Glauber Rocha. (1962).
Bolivianos. Diretor. Kiko Goifmann. Sesctv, dezembro de 2012.
Braços Cruzados Máquinas Paradas. Diretores: Roberto Gervitz e Sergio Toledo (1978).
Capitalismo – uma estória de amor. Michael Moore, 2009.
China Blue. Direção de Micha X. Peled. Estados Unidos: Teddy Bear. 2005. Dvd.
China Blue. Direção: Micha X. Peled. 2005.
Cidadão Boilesen. Diretor: Chaim Litewski. (2009).

Conterrâneos velhos de guerra. Direção de Vladimir Carvalho. Rio de Janeiro: Vertovisão, 1984. Videocassete.
Conterrâneos velhos de guerra. Direção: Vladimir Carvalho. Rio de Janeiro: 1984.
Conterrâneos velhos de guerra. Vladimir Carvalho, 1984.
Deus e o Diabo na terra do Sol. Diretor: Glauber Rocha. (1964).
Diamante de sangue. Direção de Edward Zwick, 2006.
Entre os muros da escola. Laurent Cantet, 2006.
Entrevista com Daniel Becker. www.canalsaude.fiocruz.com.br . TVNBR, 12/02/2013.
Escola Nacional Florestan Fernandes.
Ilha das Flores. Direção: Jorge Furtado. Porto Alegre: 1989.
Macunaíma. Diretor. Joaquim Pedro de Andrade. (1969).
Me matam se não trabalho e se trabalho me matam. Direção de Raymundo Gleyzer. Buenos Aires, 1974. Videocassete.
Notícias de uma guerra particular. Diretor: João Moreira Salles, 2002.
O Som ao redor. Diretor: Kleber Mendonça. 2012.
O Veneno Está Na Mesa II. Rio de Janeiro, Caliban, 2014 (Diretor Silvio Tendler).
O veneno está na mesa. Diretor: Silvio Tendler. Ano: 2008.
O Veneno Está Na Mesa. Rio de Janeiro, Caliban, 2010 (Diretor Silvio Tendler).
Os companheiros. Direção de Mario Monicelli. Roma: 1964. Videocassete.
Os companheiros. Direção: Mario Monicelli. Roma: 1964.
Os Fuzis. Diretor: Ruy Guerra. (1964).
Ou tudo ou nada. Diretor: Peter Cattaneo. Londres: 1998.
Precários inflexíveis. Diretor: Giovanni Alves, 2012.
Pro dia nascer feliz. Diretor: João Jardim, 2005.
Rio 40 graus. Diretor: Nelson Pereira dos Santos. (1957).
Segunda feira ao sol. Direção: Fernando Leon de Aranoa. Madrid: 2002.
Sickso – SOS Saúde. Michael Moore, 2006.
Terra em Transe. Diretor: Glauber Rocha. (1967).
Vidas Secas. Diretor: Nelson Pereira dos Santos. (1963)

Index

Agrochemicals 47, 167, 170

Brazil 50, 57, 58, 59, 64, 65, 66, 67, 70, 75, 77, 78, 80, 81, 83, 84, 85, 86, 87, 88, 89 92, 96, 100, 113, 114, 116, 122, 125, 141, 145, 146, 149, 150, 167, 168, 170, 173, 180, 185, 186, 187, 189, 190, 200, 202, 205, 208, 209, 211, 216, 217, 218, 219

Colonialism 8, 21, 29, 37, 38, 66, 166
Cooperation 6, 7, 9, 69, 83, 87, 94, 97, 119, 120, 139, 146, 158, 170, 175, 178, 183, 185, 187, 188, 189, 194, 195, 196, 197, 198, 212
Cooperatives 9, 96, 97, 98, 119, 178, 183, 186, 188, 189, 190, 193, 194, 198

Education 1, 2, 4, 5, 6, 7, 9, 32, 36, 44, 69, 78, 78, 88, 94, 97, 99, 114, 126, 128, 129, 137, 141, 143, 145, 150, 151, 152, 153, 154, 164, 167, 168, 171, 174, 175, 18, 182, 200, 202, 203, 204, 206, 207, 209, 210, 211, 212
Education beyond capital 69, 150, 177, 201, 211, 212, 216
Ecological agriculture 48, 50, 53, 54, 106
Ecosocialism 1, 8, 56, 61, 64, 69, 70, 73, 76, 150, 215

Green Revolution 3, 7, 8, 9, 24, 31, 32, 33, 36, 37, 46, 51, 67, 72, 77, 78, 79, 81, 82, 84, 88, 89, 90, 92, 93, 100

Imperialism 66, 85, 102, 142, 200
István Mészáros 1, 6, 54, 57, 61, 62, 63, 64, 68, 69, 70, 71, 72, 73, 75, 76, 124, 133, 134, 140, 141, 144, 147, 150, 167, 168, 169, 171, 209, 213

Karl Marx 11, 12, 13, 14, 15, 16, 17, 18, 19, 20, 21, 23, 24, 57, 62, 68, 69, 102, 119, 133, 142, 172, 201, 203, 204, 206, 210, 213

Landless Movement 153, 163, 181, 208, 215
Latin America 5, 36, 85, 86, 114, 118, 148, 151, 168, 197, 216

Michael Lowy 68, 69, 70
Miguel Altieri 8, 32, 37, 42, 43, 44, 45, 46, 47, 51, 52, 53, 106, 107, 109, 115, 117
Monoculture 7, 28, 35, 36, 37, 42, 47, 53, 74, 79, 80, 81, 90, 103, 106, 116, 119, 119, 123, 152, 194
Moisey Pistrak 129, 154, 201, 203, 205, 206, 207, 209, 213

Nadezhda Krupskaya 201, 203, 205, 206, 207, 208, 209

Primitive Accumulation 4, 7, 11, 12, 13, 14, 15, 19, 20, 21, 23, 25, 29, 164, 211, 215
Pesticides 2, 7, 24, 33, 36, 40, 41, 48, 57, 61, 67, 74, 81, 90, 103, 110, 119

Self-organization 7, 9, 128, 154, 159, 160, 162, 213
Sustainable Development 37, 50, 54, 56, 58, 59, 61, 64, 69, 75, 78, 115, 144, 173, 177, 182, 194, 202

Transnational Corporations 2, 5, 8, 9, 24, 25, 28, 61, 69, 73, 74, 122, 141, 142, 164, 165, 182, 216

Vandana Shiva 29, 32, 37, 38

www.ingramcontent.com/pod-product-compliance
Lightning Source LLC
Chambersburg PA
CBHW070616030426
42337CB00020B/3824